DIRECT SYNTHESIS OF COORDINATION
AND ORGANOMETALLIC COMPOUNDS

DIRECT SYNTHESIS OF COORDINATION AND ORGANOMETALLIC COMPOUNDS

EDITED BY

ALEXANDER D. GARNOVSKII
Rostov State University, Russia
BORIS I. KHARISOV
Universidad Autónoma de Nuevo León, México

CONTRIBUTORS

A.D. GARNOVSKII, B.I. KHARISOV, V.V. SKOPENKO,
L.M. BLANCO JEREZ, V.N. KOKOZAY, A.S. KUZHAROV,
D.A. GARNOVSKII, O.YU. VASSILYEVA, A.S. BURLOV,
V.A. PAVLENKO

1999
ELSEVIER
AMSTERDAM-LAUSANNE-NEW YORK-OXFORD-SHANNON-SINGAPORE-TOKYO

ELSEVIER SCIENCE S.A.
Avenue de la Gare 50
1001 Lausanne 1, Switzerland

First Edition 1999

Library of Congress Cataloging in Publication Data
A catalog record from the Library of Congress has been applied for.

ISBN: 0 444 72000 6

♾ The paper used in this publication meets the requirements of ANSI/NISO Z39.48-1992 (Permanence of Paper).

Printed in The Netherlands.

Contents

Preface

Conventional methods of synthesis of metal complexes as a rule involve use of metal salts or carbonyls [1–3]. The final coordination or organometallic compounds are obtained as a result of a direct interaction of components (ligand and metal source), as well as by the exchange of ligands [4] or metals [5], and using template synthesis [4]. Meanwhile, the possibility of obtaining compounds directly from compact metals by the action of an electrical current [6] was demonstrated at the end of the 19th century. Later it was reported that the synthesis of metal complexes from metals may be achieved in the gaseous phase (metals react in the form of atoms, ions or small particles) [7,8], in solutions in the presence of oxidants or in donor–acceptor systems [9,10], and under surface friction conditions [11].

The above achievements made it possible to introduce a new route in synthetic inorganic chemistry—the so-called "direct synthesis" of coordination and organometallic compounds. This "one-step" synthesis of metal complexes starting from *zero-valent* (more exactly, *elemental*) metals is nowadays an active research field which has undergone especially rapid progress over the last 25 years (12–17). Contributions to this field are of interest not only to inorganic, organic and organometallic chemists, but also to scientists and technologists involved in such diverse areas as electrochemistry, cryochemistry, mechanochemistry, sonochemistry, laser applications, matrix isolation, corrosion and catalysis.

The main advantage of "direct" reactions is the possibility of producing, on the basis of such reactions, a series of complexes that are accessible only with difficulty in conventional conditions. The specificity of the direct synthesis conditions, namely, the formation of a metal coordination sphere at the moment of complex formation, allows one to obtain new compounds readily which are unavailable by traditional synthetic routes [16].

The compounds obtained could have non-standard compositions and unusual ligand environments. Thus, pyridine $\eta^6(\pi)$-complexes could be obtained by using cryosynthesis, due to the π-coordinated ligand system of the heteroaromatic fragment [16]. A considerable number of metal π-complexes with various ligands have been obtained precisely by co-condensation reactions [17]. At present, processes under matrix isolation conditions are being studied intensively [18].

Electrosynthesis considerably increases the ability of the ligand (LH_n) protons, to form the chelates ML_n [12], as well as serving to produce other types of coordination and organometallic compounds (adducts, polynuclear, σ- and π- complexes). At the same time, coordination compounds containing an anion or a carbonyl group are frequently formed, based on the same ligands, in the conditions of the conventional syntheses (from metal salts or carbonyls). The interaction between the bulk metals or their oxides and the solutions of ligands without additional activation also gives the possibility of producing various coordination compounds having unusual structural characteristics [10]. In addition, the ultrasonic treatment of metals is frequently used for their activation [19]. Use of microwave-induced solid-state reactions involving metal powders leads to production of various inorganic compounds or refractory materials such as ceramics, silicides, carbides, chalcogenides, chlorides etc. [20,21], as well as organometallic compounds [22].

This monograph will be useful for chemists working in the areas mentioned above, especially in electrochemistry, cryoprocesses and transition metal chemistry. The material is presented in the form of generalized tables and experimental procedures which could be valuable for research and education processes.

REFERENCES

1. Inorganic Synthesis, Vols 1–24. New York: McGraw Hill (1939–1986).
2. Handbuch der Preparativen Anorganishe Chemie (Ed.: Brauer, G.) Vol. B3. Stuttgart: Ferdinand Enke Verlag (1981).
3. Comprehensive Coordination Chemistry (Ed.: Wilkinson, G.), Vols 3–5. Oxford: Pergamon Press (1987).
4. Candin, J.P.; Thaylor, K.A. *Reactions of Transition Metal Complexes*. Amsterdam: Elsevier (1968).
5. Preparative Inorganic Reactions (Ed.: Jolly, W.L.). New York: John Wiley (1964).
6. Gerdes, B. *J. Prakt. Chem.* **26**, 257 (1882).
7. Timms, P.L. *Adv. Inorg. Chem. Radiochem.* **14**, 121 (1972).
8. Klabunde, K.J. *Acc. Chem. Res.* **8**(12), 393 (1975).
9. Garnovskii, A.D.; Ryabukhin, Yu.I.; Kuzharov, A.S. *Koord. Khim.* **10**(8), 1011 (1984).

10. Lavrentiev, I.P.; Khidekel, M.A. *Russ. Chem. Rev.* **52**(4), 337 (1983).
11. Kuzharov, A.S.; Garnovskii, A.D.; Kutkov, A.A. *Zh. Obshch. Khim.* **49**(4), 861 (1979).
12. Chakravorti, M.C.; Subrahmanyam, G.V.P. *Coord. Chem. Rev.* **135/136**(1), 65 (1994).
13. Garnovskii, A.D.; Kharisov, B.I.; Gójon-Zorrilla, G.; Garnovskii, D.A. *Russ. Chem. Rev.* **64**(3), 201 (1995).
14. a) Gójon-Zorrilla, G.; Kharisov, B.I.; Garnovskii, A.D. *Rev. Soc. Quím. Méx.* **40**(3), 131 (1996); b) Kharisov, B.I.; Garnovskii, A.D.; Gójon-Zorrilla, G.; Berdonosov, S.S. *Rev. Soc. Quím. Méx.* **40**(4), 173 (1996).
15. a) Kukushkin, V.Yu.; Kukushkin, Yu.N. *Theory and Practice of Coordination Compounds Synthesis*. Leningrad: Nauka (1990), pp. 127–141; b) Davies, J.A.; Hockensmith, C.M.; Kukushkin, V.Yu.; Kukushkin, Yu.N. *Synthetic Coordination Chemistry. Theory and Practice*. Singapore: World Scientific Publishing (1996), Chapter 7.
16. *Direct Synthesis of Coordination Compounds* (Ed. Skopenko, V.V.). Kiev: Ventury (1997), 176 pp.
17. a) Klabunde, K.J. *Chemistry of Free Atoms and Particles*. New York: Academic Press (1980); b) K.J. Klabunde. *Free Atoms, Clusters, and Nanoscale Particles*. San Diego, CA: Academic Press (1994).
18. Klotzbuecher, W.E.; Petrukhina, M.A.; Sergeev, G.B. *J. Phys. Chem. A* **101**, 4548 (1997).
19. a) *Ultrasound: its Chemical, Physical, and Biological Effects* (Ed.: Suslick, K.S.). Weinheim: VCH (1988); b) Mason, T.J. *Advances in Sonochemistry*, Vol. 1. London: JAI Press (1990); c) Cintas, P. *Activated Metals in Organic Synthesis*. Boca Raton: CRC Press (1993), pp. 61–70; d) Luche, J.L.; Cintas, P. Ultrasound-induced activation of metals: principles and applications in organic synthesis. In: *Active Metals* (Ed.: Fürstner, A.). Weinheim: VCH (1996), pp. 133–190.
20. Whittaker, A.; Gavin, M.D. *J. Chem. Soc., Dalton Trans.* No. 12, 2073 (1995).
21. Hassine, N.A.; Binner, J.G.P.; Cross, T.E. *Int. J. Refract. Met.* **13**(6), 353 (1995).
22. Dabrimanesh, Q.; Roberts, R.M.G. *J. Organomet. Chem.* **542**(1), 99 (1997).

Chapter 1

Cryosynthesis of Metal Complexes

1.1 INTRODUCTION

The use of metal vapors in organic [1–4] and inorganic [4] chemistry has led to the creation of a new area of synthetic coordination chemistry: the "direct synthesis" of metal complexes starting from metal vapors in the gas phase. In addition to hundreds of original publications, a series of reviews [5–21] and monographs [22–27] have reported various aspects of this subject. Nevertheless, in the majority of these publications only particular aspects of the problem have been examined, shedding insufficient light on the overall role of direct metal-vapor synthesis in the modern chemistry of coordination and organometallic compounds. We hope that this chapter can contribute significantly to the elucidation of such a role; the aim is the metal-vapor syntheses (i.e. cryosynthesis) of the main types of coordination compounds: π-complexes and metal chelates. Indeed, as will be shown below, cryosyntheses have made a large contribution to the development of the coordination chemistry of metal σ- and π-complexes, in particular the metal-containing derivatives of olefins, alkynes their cyclic counterparts, and (hetero)aromatic systems.

There is a difference between this metal-vapor/cryosynthesis method and other types of "direct synthesis" [27] (i.e. oxidative dissolution of bulk metals, electro- and mechanosynthesis). Thus, in this case, the bulk metal must be vaporized before its reaction with gaseous or frozen (in)organic ligand. This step is necessary to provide the absence for the metal of the kinetic or thermodynamic barriers that exist for the bulk metals, and this is precisely the reason for the success of cryosynthesis.

A gaseous atom of any element except the noble gases may be expected to be more reactive than the normal form of the element for two reasons [5].

First, the atom can react faster because it has minimal steric requirements and generally has readily available electrons or orbitals. Second, the atom is a species of higher energy than the normal state of the element (for the heats of formation of the elements, see [5], p. 123).

Condensation of the atoms in an isolated form on an inert surface at 77 K will not greatly change its energy relative to room temperature, so the values $\Delta H_{298\,K}$ can be used when considering the energetics of low-temperature condensation reactions. The extra energy possessed by the atoms compared with the normal states of the metals will always make some difference to their behavior [5].

The reactions between positive or even negative metal ions and (in)organic ligands have not been included in this review: this special area has been developing rapidly and is covered in detail in several reviews elsewhere [28,29].

1.2 METHODS, CONDITIONS, EQUIPMENT, REAGENTS

Coordination compounds could be obtained in the gas phase by direct interaction of vaporized atomic metals and ligands [1–3,22,30–32]. However, the scope of this approach is strongly limited by the instability of complexes at high temperatures, which frequently leads to the breakdown of coordination compounds and/or transformation of ligands. The highest efficiency in gas-phase syntheses is attained by co-condensation of metal vapors and ligands at low temperature ("cryosynthesis") [5–8,18,22,23].

Working temperatures are usually within the interval from 10 to 273 K, although in some cases higher (295–325 K) [33–35] or lower (for example, liquid helium) [36] temperatures are used. The cryoscopic effect is examined in some reviews [7,17,18] and a monograph [23]. It is emphasized [18] that cryosynthesis using metal vapors is essentially irreversible due to the experimental conditions. The metal vapor, obtained at high temperatures (2000–2500 K) in equilibrium with the condensed phase, is transported in vacuum as a flow (current) of atoms, reacting at low temperature with the condensed phase of the ligand or its precursor (in the "matrix" of the ligand).

In order to evaluate the extent of isolation of metal atoms in the matrix of (in)organic substance, it is possible to use the ratio M/R, where M and R are the numbers of atoms (or molecules) of the matrix substance and of the metal atoms in this matrix, respectively. For the matrix, which is inert to metal [37], good isolation of atomic metals is attained when $M/R > 300$. The magnitudes $M/R = 6$–100 are usually used in cryosynthesis, since a large quantity of an (in)organic substance is evidently not necessary, due to its interaction with the atomic metal.

The matrix becomes more "rigid" when the walls of the reaction chamber are greatly cooled (to liquid-helium temperature); the diffusion of atomic atoms and their aggregation are hindered. Moreover, at this temperature it is possible to register spectroscopically the formation of compounds which are unstable at liquid-nitrogen temperature.

The equipment for direct synthesis of metal complexes in the gas phase is described in general in a series of papers [7,12,25,38–40]; its design is not complicated, since it is usually a high-vacuum system. The types of apparatus can differ according to the methods of metal evaporation used and of the introduction of the ligand into the reaction chamber. All types of modern cryosynthesis equipment (in particular, macroscale stationary co-condensation apparatus for investigations of metal atom chemistry, solution metal atom reactors, apparatus for production of gas-phase clusters, ionized cluster beam apparatus, reactors for use with volatile ligands such as CO or NO, etc.) are described by Klabunde [25]; some of these are presented in Figs. 1.1–1.4.

Metal vaporization is carried out by the following methods [5,22,25,41]: resistive heating, induction heating, bombardment with electrons of a few kilovolts, cathodic sputtering, and laser irradiation. The technological breakthroughs in metal vaporization make it possible to obtain almost all the metals of the Periodic Table in the vapor state [5,7,13,22], in particular such refractory ones as niobium, molybdenum and tungsten [2,3,7,13,15,32,42,43a]. The advantages and disadvantages of these methods of vaporization are examined in [6,7,25,41], as well as the ways of introducing the ligand into the reaction chamber and isolating the coordination compounds formed in the cryosynthesis reactions. Table 1.1 taken from a review [7], contains information about the vaporization temperatures of some metals in vacuum. The temperatures given are those at which the metal has a vapor pressure of 0.01–0.1 Torr, sufficient to give rapid evaporation under high vacuum [7].

Various organic and inorganic ligands are used in cryosynthesis, mainly unsaturated aliphatic, aromatic and heteroaromatic hydrocarbons and their functionalized derivatives, dinitrogen, dioxygen, carbon and nitrogen oxides

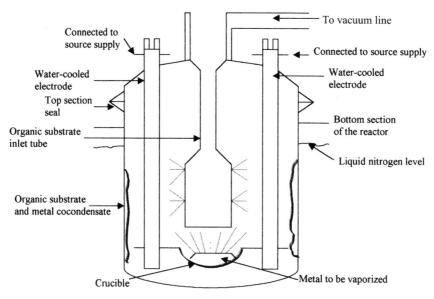

Fig. 1.1. Macroscale stationary co-condensation apparatus for investigations of metal-atom chemistry. From [25a], with permission.

TABLE 1.1
Evaporation temperatures of metals in vacuum [7]

Evaporation temperature °C					
< 1000	1000–1400	1400–1700	1700–2000	2000–2500	> 2500
Li	Be				
Na, Mg	Mn, Cu	Cr, Fe, Co	Ti, V		
		Ni			
K, Ca, Zn	Ag, Sn	Pd		Zr, Ru, Rh	Nb, Mo
Pb, Cr, Cd	Au		U	Pt	Hf, Ta
	Pr, Nd, Gd, Tb	La, Ce, Lu			W, Re
Sm, Eu	Dy, Ho, Er, Tm				Os, Ir

etc. Taking into consideration the maximum possible number of electrons which the ligand can donate to a metal, it is worthwhile to distinguish, among unsaturated hydrocarbons, monoolefins [22,33–36], diolefins [13,22,41,44–46], cyclic dienes [22,33–36,47–50] and trienes [50].

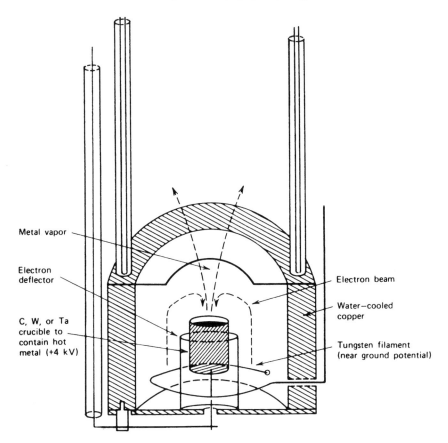

Metal vapor

Electron
deflector

C, W, or Ta
crucible to
contain hot
metal (+4 kV)

Electron beam

Water–cooled
copper

Tungsten filament
(near ground potential)

Fig. 1.2. Schematic for electron beam vaporization. From [25a], with permission.

Acetylene and its substituted derivatives have also been used as ligand systems, although less commonly [5,13,14,22,33,46]. The most widely used ligands in cryosynthesis of metal complexes are the aromatic compounds (especially benzene and its derivatives) [5,12,42,43,51–56]; heteroaromatic ligand systems include thiophene [18], pyridine and its derivatives [7,16,22,57], and α, α'-bipyridyl [58].

In addition, the following oxygen-containing organic derivatives are also used: ethers [22], ketones (in particular, acetone) [45,59], β-diketones [60], organic acids [16], their anhydrides [16] and acyl halides [13], as well as inorganic compounds [7,15].

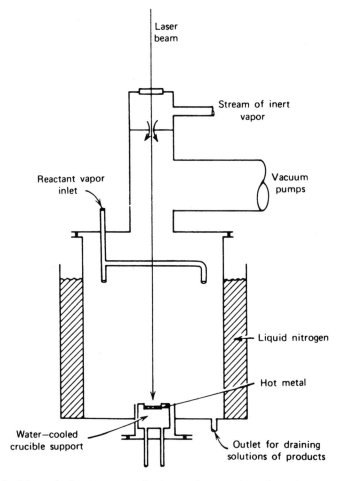

Fig. 1.3. Schematic for laser vaporization and co-condensation of vapor with chemical substrates. From [25a], with permission.

The transformations which take place during cryosynthesis of metal complexes are of two types: σ- and π-coordination of metal atoms; and insertion of metals into a C–X bond (X = H, Hal). The first type of reactions leads, mainly, to π-complexes. The second category includes the insertion of a metal into C–H [15,18] or C–Hal [15,22,30,61] bonds and the formation of organometallic compounds with σ-metal–carbon bonds.

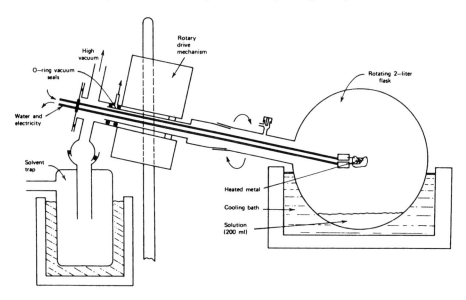

Fig. 1.4. Design of a solution metal-atom reactor. From [25a], with permission.

1.3. SYNTHESIS OF METAL COMPLEXES WITH SIMPLE INORGANIC LIGANDS

Timms [3] used phosphorus trifluoride as a ligand in one of his first pioneering studies devoted to cryosynthesis; the reactions can be presented as in equation (1.1).

$$M + nPF_3 \rightarrow M(PF_3)_n \qquad (1.1)$$

$$\mathbf{I}$$

where $M = Ni$, $n = 4$; $M = Fe$, $n = 5$; $M = Cr$, $n = 6$. The binuclear complex $Co_2(PF_3)_8$ has also been obtained according to this scheme. The same ligand has been used to synthesize other trifluorophosphine complexes [62]. All products were isolated at 77 K in yields (%) of 25 (Fe), 50 (Co), 65 (Cr), 70 (Pd) and practically 100 (Ni).

The PF_2 group has a bridging function in the bridged binuclear complexes of cobalt [3] and iron [63]. A series of complexes of other derivatives of phosphine has been obtained under the same synthetic conditions [63], for example $M(PMe_3)_n$, $M = Co$, Ni, Pd, $n = 4$; $M = Fe$, $n = 5$. The mixed-halide nickel complex of difluoromonochlorophosphine $Ni(PF_2Cl)_4$ was described in the same paper [63]. The attempt to obtain an analogous complex of phosphine itself was unsuccessful, but the mixed-ligand complex $Ni(PH_3)_2(PF_3)_2$ was isolated; it can be transformed to $Ni(PF_3)_4$ at 273 K [61].

In many cases cryosynthesis has been used to obtain the complexes of homo- and heteronuclear diatomic molecules. Thus, mono- and bis(dinitrogen)platinum(0) complexes were prepared by co-condensation of platinum vapor with N_2 molecules at 4.2–10 K in solid argon, while the reaction in solid nitrogen leads to tri- or dinitrogenplatinum(0) [64]. These compounds as well as the titanium complex $Ti(N_2)_6$ obtained by co-condensation of atomic titanium with N_2 at 10–15 K [65], were characterized by IR spectroscopy. In general, the structures of metal–dinitrogen complexes could be presented by **II–V** [66].

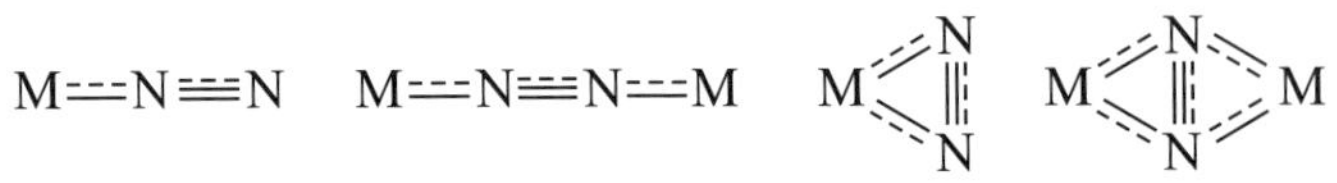

$$M\text{---}N\text{≡}N \qquad M\text{---}N\text{≡}N\text{---}M$$

II　　　　　　**III**　　　　　**IV**　　　**V**

To date, in the majority of dinitrogen complexes the metal is in the lowest oxidation state, and back-bonding, though weak, is important for the stabilization of the complexes [66].

Electron paramagnetic resonance (EPR), Fourier-transform infrared (FTIR) and UV/visible spectroscopy methods (experimental and calculated) have been used to characterize binary dioxygen complexes $M(O_2)$ and $M(O_2)_2$ of nickel, palladium and platinum in low-temperature matrices, synthesized by co-condensation of metal vapors with molecular oxygen at 4.2–10 K [67]. The mixed-ligand complexes of the type $(O_2)M(N_2)_n$, where M = Ni, Pd, Pt, were obtained in an argon matrix at 6–10 K [68]. These compounds have various compositions and structures. For example, the formulae **VI** and **VII** are assigned to the complexes $(O_2)Ni(N_2)$ and $(O_2)Ni(N_2)_2$, respectively, on the basis of the spectroscopy data.

VI　　　　　　　　**VII**

A bidentate coordination of O_2, as well as a monodentate coordination of N_2, corresponds to general ideas about the coordination of homodiatomic molecules [69].

A long series of publications, among them [70–73], are devoted to a well-known area: interaction of metal atoms with carbon monoxide. These reactions are usually conducted in a rotating cryostat which is simply a stainless steel drum, containing liquid nitrogen and rotating ($\sim$2400 rpm) under high vacuum (less than 10^{-5} Torr). Vapors of both metal and substrate are directed onto the outer surface of the rotating drum from opposite sides [74].

It has been shown that Cu atoms react with CO in a rotating cryostat at 77 K, forming CuCO and $Cu(CO)_3$, which have been identified by EPR spectroscopy [72]. CuCO is unstable and disappears rapidly above 77 K; paramagnetic $Cu(CO)_3$ is significantly more stable. In addition to these mononuclear carbonyls, the diamagnetic dinuclear carbonyl $Cu_2(CO)_6$ is formed in significant yields in adamantane and cyclohexane matrices at 77 K, indicating mobility of copper atoms on the solid hydrocarbon surface. This species is most likely produced by reaction of $Cu(CO)_3$ with a mobile copper atom followed by further reaction with CO (equation (1.2)), rather than dimerization of $Cu(CO)_3$ (equation (1.3)) [33]:

$$Cu(CO)_3 + Cu \rightarrow Cu_2(CO)_3 \xrightarrow{CO} Cu_2(CO)_6 \qquad (1.2)$$

$$2Cu(CO)_3 \rightarrow Cu_2(CO)_6 \qquad (1.3)$$

A series of metal carbonyls $M(CO)_n$, where M = Co, Mn, Cr, Fe, Ni, Pd, Pt, Rh, Cu, Ag, Ir, Eu, Nd, have been obtained in argon (or nitrogen [75], M = Cu) matrices ([21] and references therein) and studied by UV/visible spectroscopy. The synthesis of aluminum carbonyl $Al_x(CO)_2$ [16] is especially interesting, because it gives the possibility of synthesizing complexes of non-transition metals, in addition to compounds of transition metals and lanthanides. The application of Raman spectroscopy has allowed access to formation in an argon matrix (4.2–10 K) of the binary complexes $Ni(N_2)_m(CO)_{4-m}$, where $m = 1;3$, by co-condensation of nickel vapor, nitrogen and carbon monoxide [76].

The accepted model for bonding between CO and a metal is σ-donation of the p-electrons of the C to the metal orbitals with simultaneous π-back-bonding of the metal d-electrons to the unoccupied π-antibonding orbitals of CO [77]. The structures of metal carbonyls can be explained using the 18-electron rule [38,66]. It is obeyed if Cr, Fe and Ni bind six, five and four CO ligands, respectively, to give monomeric compounds. The carbonyls of Mn and Co form metal–metal bonded dimers to attain an 18-valent electron configuration. The paramagnetic 17-valent electron complex $V(CO)_6$ is an exception: in order to achieve an electron count of 18, it would have to dimerize to give a seven-coordinate species; such an arrangement is however sterically unfavorable [66]. The various structures of metal carbonyls (and isocarbonyls) are presented in [66], as well as an MO description of the metal–CO binding in them. It is emphasized that the CO ligand acts as a strong "π-acid", forming the products where a back-bonding effect takes place, stabilizing the M–CO bond. In multinuclear complexes CO can adopt doubly and triply bridged coordination modes, recognizable in neutral carbonyl complexes by characteristic $\nu(CO)$ frequencies [66]. The most recent data on metal carbonyls are presented in a review [78].

It is necessary to mention that the well-known nickel tetracarbonyl, obtained more than 100 years ago by the reaction between the bulk metal and CO, can also be synthesized by the unusual direct interaction (reaction (1.4)) between nickel vapor and CO_2 (as well as with CO) in a yield of $\sim 10\%$ [5].

$$Ni + CO_2 \rightarrow Ni(CO)_4 + NiO + CO \qquad (1.4)$$

In addition to the above-mentioned diatomic molecules, carbon monosulfide can also react with transition metals; the formation of $Ni(CS)_4$ was confirmed by IR and mass spectroscopy methods [79].

Attempts to induce nitrogen monoxide to react with metals have been made; however, in this case only stable mixed-ligand complexes $Co(NO)(PF_3)_2$ and $Mn(NO)_3(PF_3)_3$ [5] have been obtained. The latter is transformed to $Mn(NO)_3CO$ quantitatively in a CO atmosphere (1 atm, 293 K). At the same time, comparatively recently [45] it was established that the interaction of iron vapor with NO in argon at room temperature is accompanied by the formation of the complex Fe(NO) with unknown structure.

Among triatomic inorganic ligands, carbon disulfide has been used in cryosynthesis: the complexes $Ni(CS_2)_n$, where $n = 1$–3, were obtained; their formation was detected by UV spectroscopy [80]. Many recent papers are devoted to synthesis of metal complexes with other di- and triatomic molecules using matrix isolation conditions [81–88]; among them, an FTIR and quasi-relativistic density functional theory investigation of the reaction products of laser-ablated uranium atoms with NO, NO_2 and N_2O is noteworthy [81]. The reaction kinetics of chromium atoms with simple molecules such as HCl, N_2O, Cl_2, and O_2 is presented in [89].

In the case of ammonia, together with the adduct $FeNH_3$ [45], the formation of an insertion product into the N–H bond was reported; perhaps this compound has the formula $HNiNH_2$ [25,90]. A similar transformation, observed for water, leads to the product HNiOH [25,91]. An undoubted interest is the cryosynthesis of the complex of iron with methylaminodifluorophosphine $[CH_3N(PF_2)_2]_4Fe$, the structure of which has been established by X-ray diffraction [92]. It is interesting that iron has a penta-coordinated environment: three methylaminodifluorophosphine molecules play the role of monodentate N-donor ligands and one molecule acts as bidentate P,N-donor ligand.

In the case of the simplest organic molecules, the formation of stable palladium adducts with alkanes (equation (1.5)) is especially emphasized; the structure **VIII** of a donor–acceptor π-complex, which is not in keeping with the ideas about usual π-complexes, is assigned to these adducts [34]. Additionally, a recent review [93] is devoted to the activation of C–H bonds in alkanes by free metal atoms and metal clusters, as well as to the topics of σ-coordinated dihydrogen and σ-coordinated silanes.

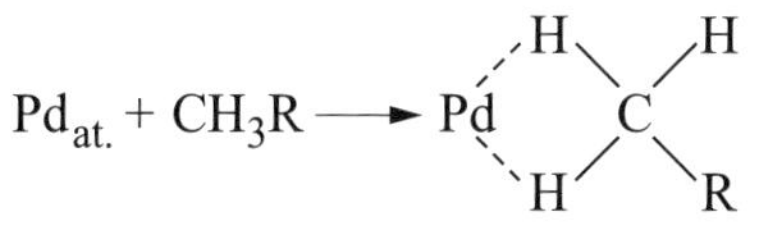

$$Pd_{at.} + CH_3R \longrightarrow Pd \begin{array}{c} H \\ \diagup \diagdown \\ \diagdown \diagup \\ H \end{array} C \begin{array}{c} H \\ \diagup \\ \diagdown \\ R \end{array}$$

$$(1.5)$$

VIII

Thus, in spite of the fact that simple (in)organic molecules were used as ligands for cryosynthesis as far back as 30 years ago [1–6], they continue to be of significant interest for studies of cryosynthetic procedures [75,78,81–89].

Table 1.2 summarizes examples of the syntheses of metal complexes with simple inorganic molecules starting from metal vapors.

TABLE 1.2
Cryosynthesis of metal complexes with simple molecules

Metal	Ligand	Product	T, K	Yield %	Ref.
Ni, Pd, Ag, Pt, Rh_n $n = 1-3$	O_2	$M_m(O_2)_i$ $m = 1-4$ $i = 1, 2, 4, 6, 8, 12$	4, 2–77	5–15	67, 94–96
Cr, Fe, Co, Ni, Pd	PF_3	$M_n(PF_3)_m$ $n = 1, 2$ $m = 3-8$	77	25–85	63, 97
Fe	PF_3	$(PF_3)_3Fe(PF_2)_2Fe(PF_3)_3$	77	25	63, 97
Ni	$PCIF_2$	$Ni(PF_2Cl)_4$	77	32	63, 97
Ni, Mn	$PF_3 + X$ $X = PH_3$, NO $(n = 2, 3;$ $m = 1, 2)$	$Ni(PF_3)_nX_m$	77	10–15	63, 97
Pt, V, Ti, Pd	N_2	$M(N_2)_n$ $n = 1, 3, 6, 12$	4–15		6, 63, 64, 98, 99
Ni, Ti, Ag, Rh, Ir, V, Al, Cu etc.	CO	$M_n(CO)_m$ $n = 1-4; m = 1-12$	4–77		5, 6, 16 5, 71–73, 76, 77, 100–106
Ni, Pd, Ag, Pt, Au, Co	$O_2 + X$ $X = N_2$, CO	$(O_2)M(X)_{1-4}$	6–12	6–40	68, 107–109
Ni	$CO + N_2/Ar$	$Ni(CO)_n(N_2)_m$ $n, m = 1-3$	4,	2–10	76
Ni	CS_2	$Ni(CS_2)_n$ $n = 1, 3$	10–12		80
Ag	SiO	$Ag(SiO)_{1-3}$	77		110
Cr	$P(OMe)_3$	$Cr\{P(OMe)_3\}_6$	77		6
Cu	CH_4	$HCuCH_3 \rightarrow CuCH_3$, CuH	12		111
Li	CCl_4 C_2Cl_6	Li_4C Li_6C_2	1073– 1273		112
Ag	CO_2	$Ag(CO_2)$	10–25		113
Cr, Mo, Co, Ni, Ag, Au, Fe, Cu, Pd	Molten salts $\{KNO_{3-}$ $LiNO_{2-3};$ (K, Na, Li)OAc; KNCS\}	Metal oxides or metal powders	413– 443		114

Experimental procedures

M + PF₃

Condensation of transition metal vapors with trifluorophosphine was caried out at 77 K. In each case the metal vapor was condensed with PF_3 in a mole ratio of at least 1:8. The following products were isolated [63,97]:

Cr:	$Cr(PF_3)_6$ (yield 65%)
Mn,Cu:	No volatile product
Fe:	$Fe(PF_3)_5$ (25%) and $(PF_3)_3Fe(PF_2)_2Fe(PF_3)_3$ (25%)
Co:	$[Co(PF_3)_4]_x$ (50%)
Pd:	$Pd(PF_3)_4$ (70%)

Cu + BCl₃.

Diboron tetrachloride was formed in yields varying from about 70% with a BCl_3Cu mole ratio of 20:1, to 40% with a 6:1 mole ratio. About 300 mmol of copper, evaporated in 50 min, were condensed with 2.7 mol of BCl_3 to form 57 mmol of B_2Cl_4 [5]:

$$2Cu(g) + 2BCl_3(g) \rightarrow 2CuCl + B_2Cl_4$$

Cu or Ag + PCl₃

Both copper and silver vapors react very readily with PCl_3 at 77 K. Some P_2Cl_4 can be pumped off from the co-condensates on warming, but most of the dechlorinated phosphorus species remain coordinated to the metal in highly colored solids [5].

$$Cu(g) + PCl_3(g) \rightarrow CuCl(PCl)_x \ (10\%) + P_2Cl_4$$

$$Ag(g) + PCl_3(g) \rightarrow AgCl(PCl)_x \ (15\%) + P_2Cl_4$$

Ni + PH₃ + PF₃

An equimolar mixture of PH_3 and PF_3 condensed with nickel at 77 K formed two volatile compounds $Ni(PF_3)_2(PH_3)_2$ and $Ni(PF_3)_2PH_3$, which were separated from $Ni(PF_3)_4$ by low-temperature distillation [63]. Some hydrogen was also evolved when the condensate was warmed from 77 K. The compound $Ni(PF_3)_2(PH_3)_2$ decomposed slowly above 273 K evolving hydrogen. When allowed to warm to room temperature in the presence of PF_3, it was converted quantitatively to $Ni(PF_3)_2PH_3$.

Mn + NO + PF₃

The compound $Mn(PF_3)(NO)_3$ was formed in 25% yield from manganese vapor, NO, BF_3 and PF_3 [63]; it could not be made using conventional high-pressure techniques.

Ni + CO₂

When nickel vapor was condensed with CO_2 at 77 K, the resulting solid evolved CO rapidly on being warmed above 123 K. About 10% of the nickel was recovered as $Ni(CO)_4$ [63]:

$$Ni(g) + CO_2(g) \rightarrow NiO + CO + Ni(CO)_4$$

Ag + O₂

Silver atoms were co-condensed with O_2 or with O_2/Ar matrices at 6 K (using very low concentrations of Ag to eliminate complications due to cluster formation). In the first case, the resulting infrared spectrum showed a closely spaced doublet at 1102/1097 cm^{-1} and a very weak absorption at 440 cm^{-1} (O–O stretching mode). It was established that there was one dioxygen molecule in the Ag^+,O_2^- complex formed from isotopic experiments with $^{16}O_2/^{18}O_2 \simeq$ 1:1 [94].

Ni + CS₂

Monoatomic nickel vapor was generated by directly heating a 0.010-in (0.25 mm) ribbon filament of the metal. Research-grade carbon disulfide (99.99%) was doubly distilled before use, using standard vacuum-line techniques. The rate of metal atom deposition was followed continuously and controlled using a quartz crystal microbalance. The metal concentration could be set anywhere in the range Ni/matrix $\simeq$ 1:10^6 to 1:10 by simply varying the metal and matrix deposition rates. Matrix gas flows, controlled by a calibrated micrometer needle valve, were maintained in the range 0.1–2.0 mmol h^{-1}. The following products were detected by UV/visible and IR spectroscopy: $Ni(CS_2)$, $Ni(CS_2)_2$ and $Ni(CS_2)_3$ [80].

Ni + N₂ + CO

Monoatomic nickel vapor was generated by directly heating a thin (0.010 in; 0.25 mm) ribbon filament of metal. Gas mixtures of CO, N_2 and Ar (1:1:20) were prepared by conventional vacuum-line techniques. The co-condensation was carried out at 10 K. Various products have been obtained: $Ni(CO)_2(N_2)_2$, $Ni(CO)_3N_2$, $Ni(CO)_2(N_2)_2$, $NiCO(N_2)_3$ etc. [76].

Metal atoms + molten salts

Metal (Cr, Mn, Fe, Co, Ni, Cu, Ag, Au, Pd) was evaporated (0.1–2 g over 1 h) from a resistance-heated crucible in a flask while the salt was kept just molten by applying flame-heat or air-cooling, depending on the heat radiation from the crucible. Condensing atoms entered the melt to give slurries or solutions of the products.

Products insoluble in the melt were isolated anaerobically, either by dissolving the solidified salts with water or other solvents and filtering, or by hot filtration of the melt through a glass frit followed by washing of the collected solid at < 100°C to remove entrained salts. Products soluble in the melt were more difficult to isolate; success depended on finding suitable selective solvents for extraction from the solid salts. The following metals and metal oxides were detected as products of co-condensation: Cr_2O_3, MnO_2, Fe_2O_3, Co_3O_4, NiO, Cu_2O; Ag, Au, Pd powders [114].

1.4 CRYOSYNTHESIS OF METAL π- AND σ-COMPLEXES

1.4.1 Interaction of metals with olefins

The simplest π-complex compounds are obtained by the interaction of metal vapors with alkenes (monoolefins) (equation (1.6)). It was shown that the co-condensation of silver vapor and ethylene in the presence of dioxygen at 10 K leads to the unstable ethylene–silver complex **IX**, which decomposes at 40 K.

$$Ag_{at.} + C_2H_4 + O_2 \xrightarrow{10\,K} \left[\begin{matrix} H_2C \\ \| \\ H_2C \end{matrix} \longrightarrow Ag \right]^+ [O_2]^- \xrightarrow{40\,K} \text{Decomp.} \qquad (1.6)$$

IX

The ethylene complexes of cobalt (in argon, 12–15 K [115]), nickel (in argon, 10–77 K [116]; in helium [36]), palladium (in helium, 300±5 K, 0.5–0.8 Torr [34]), and copper (in argon, 295 and 395 K, 500–600 Torr [33]) are described. The gas-phase kinetics of the reactions of neutral transition metal atoms with olefins (ethylene, propene, butene and isobutene) [35] was studied in the cases of the direct interaction between yttrium, molybdenum, zirconium and niobium with alkenes.

A series of platinum π-complexes with propene and butene, as well as with allene [117], have been obtained at 77 K. The effective rate constants of these reactions were determined under helium pressure (0.5–0.8 Torr) at 300 K; it was shown that in similar conditions Y, Zr and Nb interact rapidly with the above alkenes while Mo reacts slowly [35]. The kinetic data, corresponding to the interaction of monoolefins (ethylene, propene, 1-butene, 0.5–0.8 Torr) with transition $3d$- and $4d$-metals (Sc, Ti, V, Ni etc.), are presented in [34,36,45] and references therein.

The 1:1 adducts are formed [45] by the interaction of ethylene with iron, while copper forms the complex $Cu(C_2H_4)_2$ [33]. In the case of the interaction between gallium atoms [118] with ethylene in adamantane on a rotating cryostat at 77 K, a product assigned as the cyclic σ-bonded gallacyclopentane **X** is obtained.

X

There are no X-ray diffraction data in the available literature on metal complexes with olefins, obtained by cryosynthesis; however, the ideas [35] and the proposed scheme of a donor–acceptor model of the metal–alkene bond (Fig. 1.5) [38] are useful for understanding the coordination of a metal in olefin complexes. The ligand donates π-electron density to a metal orbital of σ-symmetry directed to the center of the ligand π-system, and the metal in turn back-bonds electron density into a ligand π^*-orbital. The result is a synergism that, as in the case of the metal carbonyls, leads to relatively strong bonding [38]. Even in this simple form it is clear that such bonding can occur effectively only with a low-valent metal (i.e. in oxidation states -1 to $+2$) with populated π-symmetry orbitals (i.e. a metal late in the transition series).

In the example of the Al–C_2H_4 complex, the structure of such metal complexes could be explained using Density Functional Theory (DFT) and the topological method (ELF) [119]. It is shown that this complex has a C_{2v} symmetry equivalent structure and a $^2B_2$ electronic ground state, which is strongly bound by -13.3 kcal mol^{-1} (-55.6 kJ mol^{-1}) (to be compared with an experimental value of -16 kcal mol^{-1}) (-66.9 kJ mol^{-1}); the Al–ethylene bonding is mostly electrostatic [119]. A

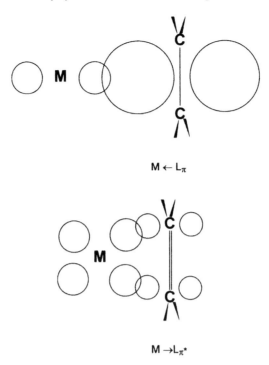

Fig. 1.5. Bonding in metal–olefin complexes.

theoretical study of the reaction of copper dimer with ethylene using SCF/CASSCF/MP2 methods is described in [120]. It was revealed that the perpendicular approach of Cu_2 is more favored than the parallel. The strong bonding is assisted through the donation of π-electronic density from ethylene to the $4p$-orbital of Cu.

Additionally to metal–alkene reactions in cryosynthesis conditions, the interaction of transition metals and halogen-substituted unsaturated hydrocarbons also leads to π-complexes. Thus atomic palladium, reacting with *cis*-perfluorobutene (1.7) at 77 K, forms the coordination compound **XI** [22].

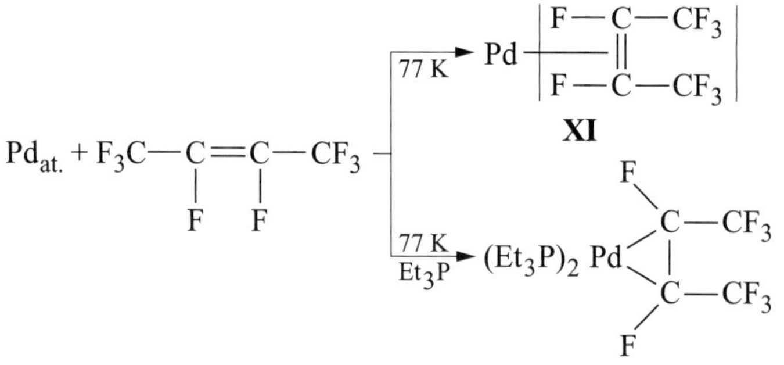

$$\text{Pd}_{\text{at.}} + \text{F}_3\text{C}-\underset{\underset{\text{F}}{|}}{\text{C}}=\underset{\underset{\text{F}}{|}}{\text{C}}-\text{CF}_3
\begin{cases}
\xrightarrow[\;\;]{77\,\text{K}} \text{Pd} \left|\begin{array}{c} \text{F}-\text{C}-\text{CF}_3 \\ \| \\ \text{F}-\text{C}-\text{CF}_3 \end{array}\right| & (1.7) \\[4pt]
\textbf{XI} & \\[10pt]
\xrightarrow[\text{Et}_3\text{P}]{77\,\text{K}} (\text{Et}_3\text{P})_2\,\text{Pd} & (1.8) \\
\textbf{XII} &
\end{cases}$$

Complex XII structure:

$$(\text{Et}_3\text{P})_2\,\text{Pd} \underset{\text{F}}{\overset{\text{F}}{\diagup \diagdown}} \begin{array}{c}\text{C}-\text{CF}_3 \\ | \\ \text{C}-\text{CF}_3\end{array}$$

The same reactants in the presence of triethylphosphine (equation (1.8)) form σ-complex **XII** [13,16]. However, the structures of complexes **XI** and **XII** cannot be confirmed on the basis of available data.

Gas-phase reactions of alkenes and their halogen-substituted derivatives with atomic metals which are accompanied by formation of π-allyl complexes are known. In particular [7,22], π-allyl complex **XIII** is formed by the interaction at 77 K of propene with atomic cobalt in the presence of phosphorus trifluoride. The reaction of bromopropene with atomic nickel at 77 K and 298 K is accompanied [22] by the formation of bisallyl derivative **XIV**.

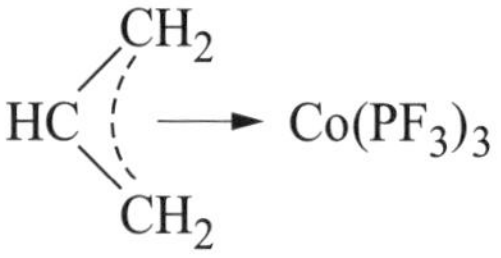

$$\textbf{XIII} \qquad\qquad\qquad \textbf{XIV}$$

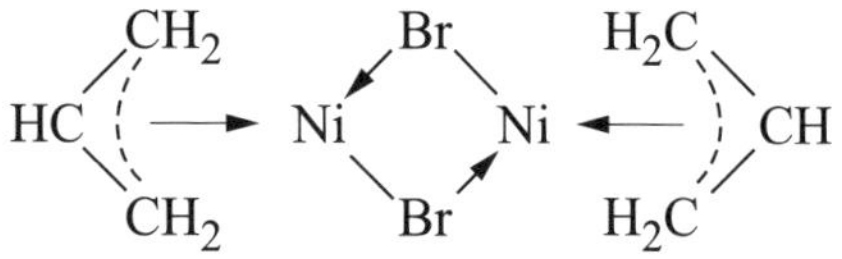

Organometallic compounds can serve as a source of allyl ligands, for example in the formation of π-allyl complexes of nickel (equation (1.9)) and chromium (equation (1.10)) complexes [22].

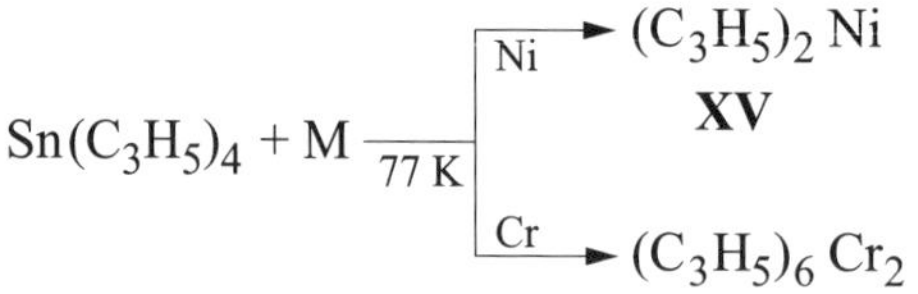

$$\text{Sn}(\text{C}_3\text{H}_5)_4 + \text{M} \xrightarrow[77\,\text{K}]{}
\begin{cases}
\xrightarrow{\text{Ni}} (\text{C}_3\text{H}_5)_2\,\text{Ni} \qquad\qquad (1.9) \\
\qquad\;\; \textbf{XV} \\[6pt]
\xrightarrow{\text{Cr}} (\text{C}_3\text{H}_5)_6\,\text{Cr}_2 \qquad\;\; (1.10) \\
\qquad\;\; \textbf{XVI}
\end{cases}$$

However, choosing the olefins as ligands for the cryosynthesis, it is necessary to bear in mind that various catalytic transformations of these unsaturated hydrocarbons could take place during their co-condensation with metal vapors.

Selected data on metal vapor reactions with alkenes are presented in Table 1.3.

Experimental procedures

Co + C$_2$H$_4$

Cobalt vapor was generated by directly heating a 0.01-in (0.25 mm) ribbon filament of the metal and co-condensed with C$_2$H$_4$/Ar matrices at 12–15 K. A series of mono- and binuclear cobalt–ethylene complexes Co(C$_2$H$_4$)$_i$ (where $i = 1$, 2) and Co$_2$(C$_2$H$_4$)$_m$ (where $m = 1$, 2) as well as a suspected tetranuclear species Co$_4$(C$_2$H$_4$)$_n$ have been detected by spectroscopic methods [115].

Ni (Pd) + olefins and their derivatives

Metals were vaporized by the same technique and co-condensed with a series of olefins at 10–77 K. The following products have been detected: Ni(C$_2$H$_4$)$_{1-3}$, Ni(C$_3$H$_6$)$_{1-3}$, Ni(C$_4$H$_8$)$_{1-3}$ (with isobutene, *cis*- and *trans*-but-2-ene), Ni(CH$_2$CHCl)$_{1-3}$, Ni(CH$_2$CHF)$_{1-3}$, Ni(CH$_2$CHCH$_2$Cl)$_{1-3}$, Ni(C$_2$ F$_4$)$_{1-3}$, Pd(C$_2$H$_4$)$_{1-3}$, Pd(C$_3$H$_6$)$_{1-3}$ etc. [116]. Similarly, the copper complexes Cu(C$_2$H$_4$)$_{1-3}$ [126] and Pd(C$_2$H$_4$) [123] have been reported.

Cu, Ag or Au + higher alkenes

The reactions were carried out in a rotating cryostat. The metal vapors were co-condensed with a range of higher acyclic and cyclic alkenes in inert hydrocarbon matrices (adamantane and cyclohexane) at 77 K. The products were examined by electron paramagnetic resonance spectroscopy. The following complexes were detected, among others [129]: Cu[*c*-C$_6$H$_8$], Cu[*c*-C$_8$H$_{14}$], Ag[*c*-C$_8$H$_{12}$], Au[*c*-C$_6$H$_{10}$].

1.4.2 Interaction of metals with polyenes

Under the conditions of metal vapor synthesis, noncyclic diene hydrocarbons form a series of π-complexes in which unsaturated ligands mainly have a role as η^4–π-donor molecules [7,11,13,16,22,41,44,45]. Most of these π-complexes are derived from 1,3-butadiene, for which five types of coordination compounds having the metal/diene compositions 1:1:0, 1:1:n, 1:2:0, 1:2:n and 1:3:0 (L = CO, PF$_3$, PR$_3$), have been prepared. According to [45], com-

TABLE 1.3
Cryosynthesis of metal complexes with olefins

Metal	Ligand	Product	T, K	Ref.
Ni, Pd, Au, Li, Co, Cu etc.	C_2H_4	$M_n(C_2H_4)_m$ $n = 1-4$; $m = 1-6$	8–77	22, 115, 121–126
Ni	$Sn(C_3H_5)_4$	$Ni(C_3H_5)_2$	77	22
Cr	$Sn(C_3H_5)_4$	$Cr_2(C_3H_5)_2$	77	22
Co	$EtCH=CH_2 +$ PF_3	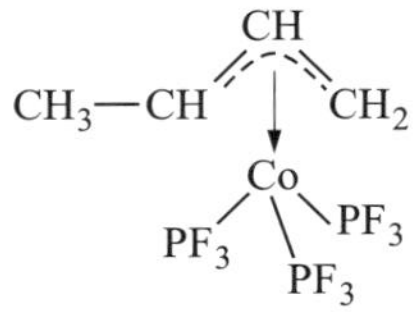	77	11, 127
Pd	Bicyclo[2.2.1]-heptene (L)	PdL_3	77	128
Li	C_2H_4	$Li(CH_2)_nLi$ $n = 2, 4, 6$	77	125
Cu, Ag	(A)cyclic alkenes (L)	ML	77	129
Al	$CH_3 - CH = CH_2$	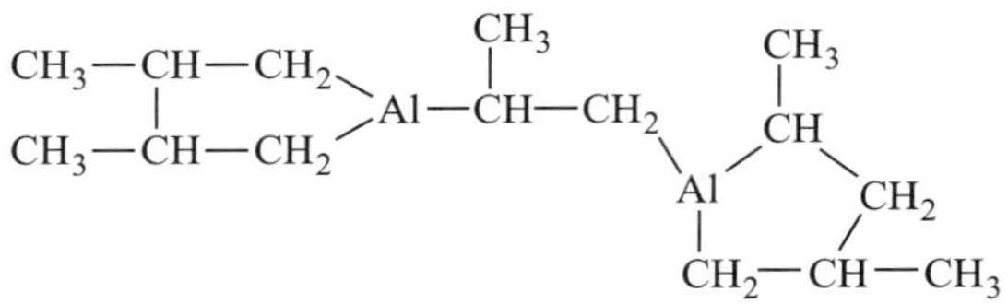	77	11, 127

plexes of the first type are formed by interaction of 1,3-butadiene with iron at room temperature and 100 Torr, those of the second type by the reactions of diene and trimethoxyphosphine [130] or triphenylphosphine [117]. The formation of the third type of coordination compounds is deemed possible in the cases of cobalt [22] and iron [107]. The fourth (and most common) type of compound includes the iron [13,41], cobalt [11,13], chromium [13,132] and ruthenium [44] complexes. In particular, ruthenium forms the stable adducts $Ru(\eta^4\text{-}C_4H_6)_2L$, where L = PF_3, CO, Me_3CNS [44]. The complexes of the fifth type, formed by molybdenum and tungsten {tris(butadiene)molybdenum (tungsten)} [11,127,133], are characterized by X-ray diffraction data, which show equal distances from the Mo and W atoms to all the carbon atoms of the butadiene ligand and a trigonal prismatic arrangement of the C=C double bonds **XVII** [38].

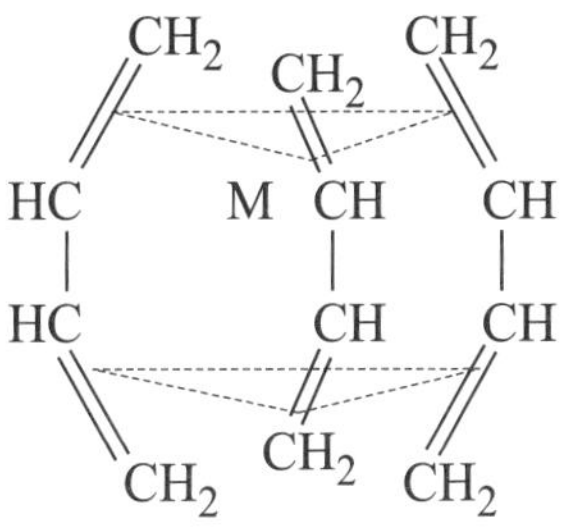

XVII

The photochemical cryosynthesis (equation (1.11)) of iron complexes with the participation of propanediene (allene) in solid argon takes place with high selectivity [46]. The formation of π-complexes of diatomic iron **XVIII** was observed by FTIR spectroscopy.

XVIII

$$(1.11)$$

The structures presented by formulae **XVIIIa** and **XVIIIb** were proposed for these π-complexes:

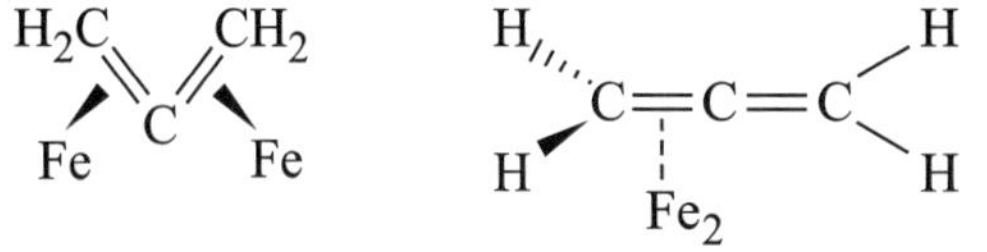

XVIIIa   **XVIIIb**

An ESR (electron spin resonance) study of the reaction of ground-state Al atoms with 1,3-butadiene in adamantane at 77 K in a rotating cryostat showed the formation of two paramagnetic products: a σ-bonded Al cyclopentene and an Al-substituted allyl [134]. The authors proposed the structure of the first complex with the Al atom below the plane of the butadiene C framework at an angle of $\sim 42°$, whereas the Al-substituted allyl is possibly a bridged species.

Cycloolefin complexes [13,22,47,50] are among the most interesting coordination compounds prepared by cryosynthesis. The compound **XIX** [22] is an example of a "nonstandard" binuclear π-complex, obtained as a result of the transformation (1.12).

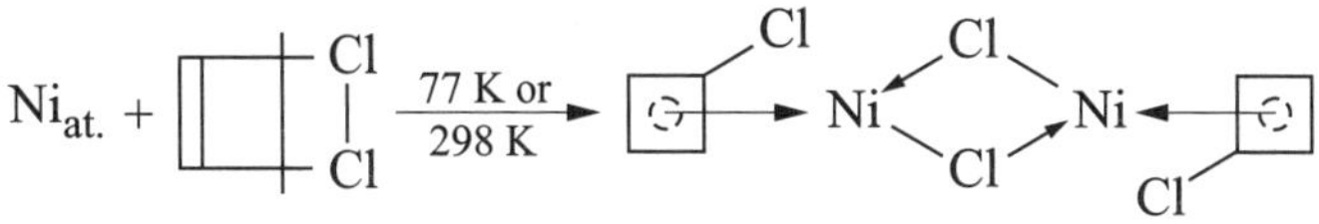

$$(1.12)$$

XIX

Most η^5–π-coordinated complexes having the general formula **XX** have been obtained starting from the basis of cyclopentadiene and its substituted derivatives [9,13,16,22,47,49,135,136], which have a very important role in metal vapor chemistry. In this case, the hydrogen atom is lost with a simultaneous aromatization of the ligand system (reaction (1.13)).

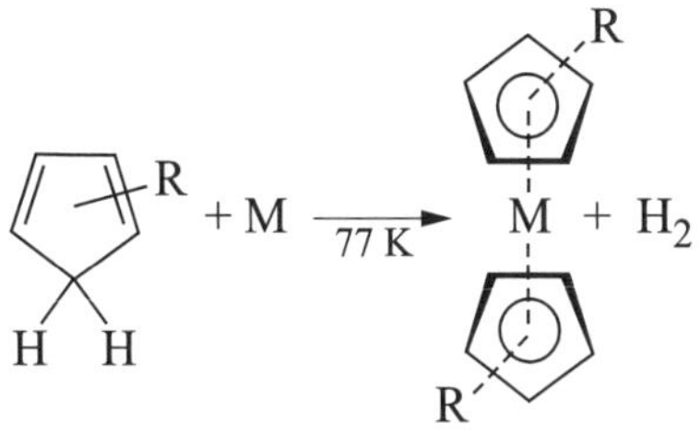

$$(1.13)$$

XX

M = Co, Fe, Cr, Mo, W

Thus, Nesmeyanov et al. [136] have developed synthesis conditions for symmetrically substituted heteroannular alkylferrocenes starting from alkylcyclopentadienes (ligand/metal molar ratio = 20–40), which were co-condensed on the surface of the reactor at 77 K with iron vapor. The reactions took place according to Eq. (1.13), where $R = Alk_n$, $M = Fe$. The yields of alkylferrocenes were 28% (R = dimethyl), 29% (R = octamethyl), and 72% (R = diethyl) [136].

An interesting series of transformations (scheme (1.14)) with spiro derivatives of cyclopentadiene has been described [137]. As a result of the interaction of atomic molybdenum with spiro[2.4]hepta-4,6-diene at 77 K, the cyclopentadienyl derivative **XXI** was obtained, from which a series of sandwich molybdenum complexes of general formula **XXII** could be prepared.

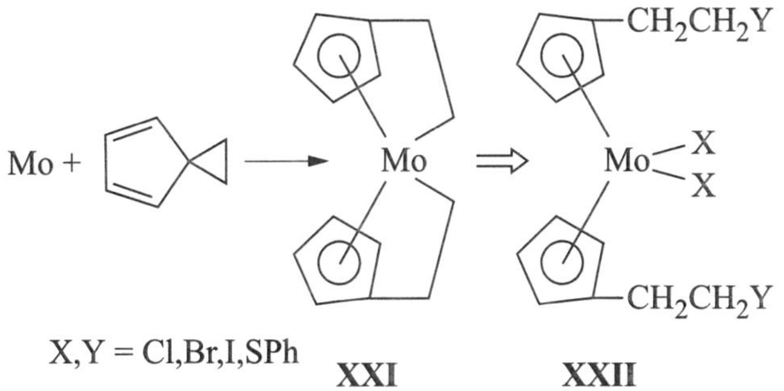

$$X,Y = Cl, Br, I, SPh \qquad (1.14)$$

X, Y = Cl,Br,I,SPh   **XXI**   **XXII**

An intermolecular dehydrogenation–hydrogenation of the cyclopentadienyl ligand takes place in reaction (1.13) (M = Ni, R = H), forming **XXIII**.

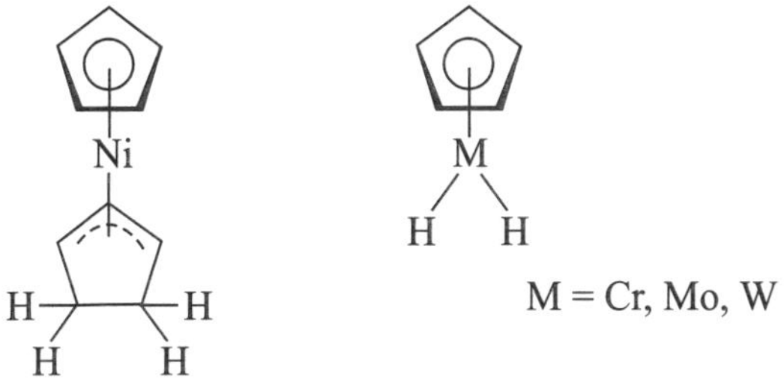

M = Cr, Mo, W

XXIII   **XXIV**

Additionally to the typical sandwich complexes, the monocyclopentadienyl–hydride complexes **XXIV** have also been isolated after interaction of cyclopentadiene with atomic chromium, molybdenum or tungsten [13].

The derivatives of cyclopentadiene (Cp) with various substituents (L = 1,3-t-Bu$_2$CpH, 1,2,4-t-Bu$_3$CpH, EtMe$_4$CpH, M = Co [138]; L = 1,3-t-Bu$_2$CpH, 1,2,4-t-Bu$_3$CpH, M = Ni [139]) were obtained by the metal vapor route. The structures of the products were determined by single-crystal X-ray diffraction. Co-condensation of metals with 6,6-dimethylfulvene and a spectroscopic study of the complexes formed have been carried out [140,141]. It was shown that Ga, Tl, Cu and Mn form Cp$_i$M(I) derivatives (Cp$_i$ = C$_5$H$_4$CHMe$_2$); when M is Cu or Mn, trimethylphosphite is needed as a co-ligand to fill up its coordination sphere. In the case of cobalt, its interaction with 6,6-dimethylfulvene and trimethylphosphite leads to the formation of CoH[P(OMe)$_3$]$_4$ instead of a half-sandwich complex [140]. Sn and Pb form *anse*-metallocenes **XXV** [141].

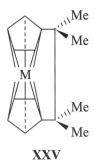

XXV

It is noteworthy that this is one of the very few examples of a Pb complex where Pb undergoes a direct synthesis to an organometallic [141].

Reaction of a structural analogue of cyclopentadiene (indene, **XXVI**) with atomic iron and tungsten leads to compounds **XXVII–XXX**, for which a peculiar bonding isomer is typical; it is caused by the participation of either cyclopentadienyl or benzene rings in the coordination with the metal atom [22]. This interesting phenomenon has been examined in a limited series of cases; the reasons provoking it are still not clear.

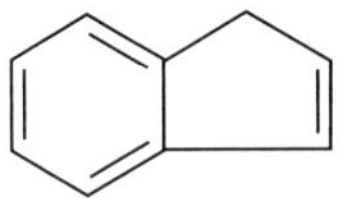

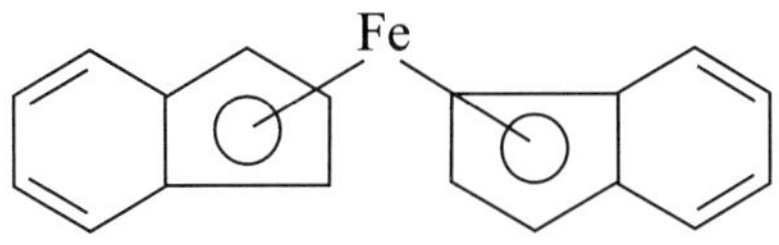

XXVI   **XXVII**

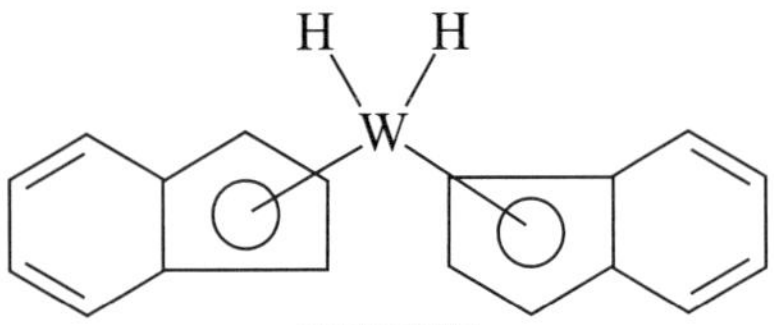

XXVIII

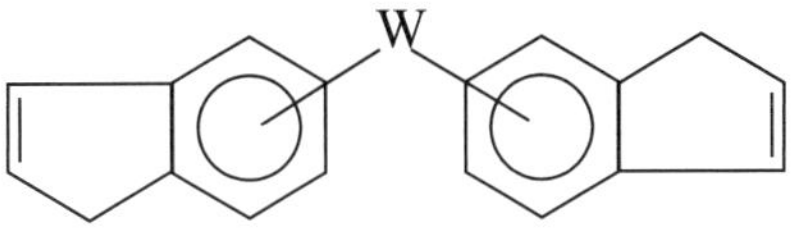

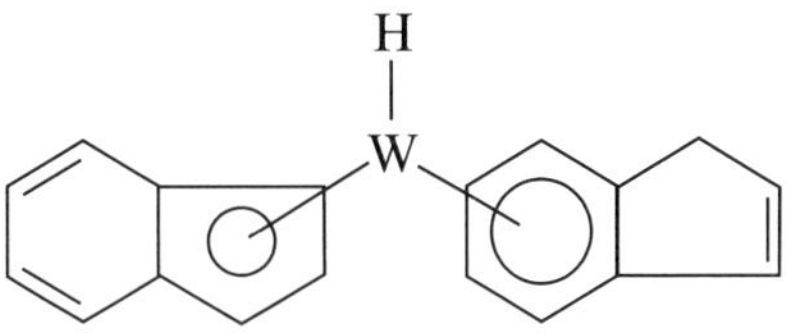

XXIX   **XXX**

Among the mixed-ligand compounds, cyclopentadienylcarbonyl $C_5H_5M(CO)_2$, cyclopentadienylpentamethylcyclopentadienyl $C_5H_5MC_5$ $(CMe_5)_5$, and cyclopentadienylacetylene $C_5H_5M(C_2H_2)_2$ complexes, where M = Rh and Ir, have been cryosynthesized and characterized [48]. The application of the metal-atom technique to the synthesis of Cp-metallaboron clusters has been investigated [142]. Thus, a series of the products was obtained by the reaction of Co vapor and cyclopentadiene with B_5H_9, including 1,2,3-$(CpCo)_3B_5H_5$ (Cp = η^5-cyclopentadienyl), 8-σ-(C_5H_9)-1,2,3-$(CpCo)_3B_5H_4$ and $(\mu_3$-CO)-1,2,3-$(CpCo)_3B_3H_3$, among others.

Cryosynthesis with participation of atomic metals allows one to produce the π-complexes of cyclohexa-1,3-diene (CHD–C_6H_8), as well as cycloocta-1,3- and -1,5- dienes (CHD-C_8H_{12}) [16,22,50]. The complex compounds formed as a result of these transformations have the compositions MLL'_n, ML_2, ML_2L', where L = diene; L′ = H, CO, PF$_3$, C_4H_9NC, $Ph_2PCH_2CH_2PPh_2$; M = Fe, Cr, Mn. The complex $Mn(\eta^4$–1,3-CHD)$_2$CO [50] has been obtained by co-condensation of atomic manganese with cyclohexa-1,3-diene and CO at 77 K with further heating of the condensed phase to 93 K. Its structure has been proved beyond reasonable doubt by X-ray diffraction [50]. Two complex compounds from manganese and cycloocta-1,3-diene, having compositions $Mn(\eta^5$-$C_8H_{11})(CO)_3$ and $Mn(\eta^3$-$C_8H_{13})$

$(CO)_4$, were isolated under the same conditions [50], through dehydrogenation–hydrogenation of the ligand system.

The details of co-condensation processes of 1,3- and 1,5-cyclooctadienes with metals were examined in a review [16] and monograph [22] in 1979, since when only one paper concerning this type of ligand has appeared [50], to our knowledge. Among the results reviewed in [16] two must be highlighted: the isomerization with chromium of 1,3-cyclooctadiene (1,3-COD) to the 1,5-diene ligand system and the X-ray diffraction proof of the structure of the complex $C_8H_{11}Cr(PF_3)H$ as (η^4-cyclooctadienyl)hydridotris(trifluorophosphine)chromium. Additionally, the complex $(COD)_2Fe$, prepared by the metal-vapor route, was used for the subsequent ligand exchange reaction of one COD ligand. The aim was to prepare (COD)(phosphine)Fe(0) complexes which could be used as novel room-temperature catalysts for pyridine formation [143].

The syntheses of a series of π-complexes of cyclotrienes [49,144] have been carried out. Thus, interaction of the vapors of titanium, vanadium, chromium and cobalt with cycloheptatriene (C_7H_8) at 77 K leads to the mixed-ligand compounds $Ti(\eta\text{-}C_7H_7)(\eta^5\text{-}C_7H_9)$, $V(C_{14}H_{16})$ and $Cr(\eta\text{-}C_7H_7)(\eta^4\text{-}C_7H_{10})$, as well as the bis(heptatriene)chromium complex [49]. However, it was impossible to isolate complexes of manganese, nickel and palladium with these ligands.

By co-condensation of chromium vapor with mixed cycloheptatriene and trifluorophosphine in the same conditions, the complex (1–6-η-cyclohepta-1,3,5 trienyl)-tri-(trifluorophosphine)chromium(0) $[Cr(C_7H_8)(PF_3)_3]$ was obtained. Its structure was established by IR spectroscopy and heteronuclear ^1H, ^{13}C and ^{31}P NMR data [144]. The η^5-cycloheptatrienyl-η^5-cyclopentadienyl iron complex $Fe(\eta^5\text{-}C_7H_7)(\eta^5\text{-}C_7H_9)$, obtained by co-condensation of iron vapor and liquid cycloheptatriene in methylcyclohexane at 143 K, was also characterized [144]. Starting from cycloheptatriene under cryosynthesis conditions (77 K), manganese complexes $Mn(\eta^4\text{-}C_7H_8)_2(CO)$ and $Mn(\eta^5\text{-}C_7H_9)(CO)_3$ were synthesized [50], although with low yields. Other types of complexes prepared from cycloheptatrienes are discussed in the reviews [7,16].

Only a few cyclooctatetraene (COT) complexes have been obtained by cryosynthesis [144–146]. The syntheses of bis(cyclooctatetraene)uranium $U(COT)_2$, analogous thorium and plutonium complexes, and (COT)-tris(trifluorophosphine)iron (COT)Fe$(PF_3)_3$ have been reviewed [21]. Although the binuclear COT complex $Cr_2(C_8H_8)_3$ has been mentioned [144], its synthesis was not carried out. A paramagnetic sandwich complex of tris(cycloocta-tetraene)dititanium has been described [146].

Summarized data on the vapor synthesis of metal complexes with dienes, trienes and tetraenes are presented in Table 1.4.

Experimental procedures

Mo + spiro[2.4]hepta-4,6-diene

Molybdenum atoms (6.0 g, 62.5 mmol), generated from a molten ingot (ca. 25 g), were co-condensed with spiro[2.4]hepta-4,6-diene (150 cm^3) onto the reactor wall for 4 h. The wall was cooled by liquid dinitrogen. The mixture was allowed to warm to room temperature and was then washed from the reaction vessel with tetrahydrofuran (THF) (1000 cm^3). The extract was filtered through a Celite bed and then the solvent and excess of spiro[2.4]hepta-4,6-diene were removed under reduced pressure. The residue was extracted with light petroleum (4 × 500 cm^3, b.p. 60–80°C) giving a dark red solution which was filtered and the filtrate was concentrated to ca. 50 cm^3 under reduced pressure. Cooling to 253 K and then to 195 K gave large red crystals which were collected, washed with cold light petroleum (2 × 30 cm^3, b.p. 30–40°C) and dried in vacuo. The yield was 10–12 g, 35–43 mmol, 50–60% [137].

W + 1,3-butadiene

Tungsten atoms were generated by resistive heating of a metal wire. Co-condensation of the atoms with 1,3-butadiene (molar ratio 1:100) at liquid-nitrogen temperature produced a yellow matrix. After warm-up the yellow-brown liquid was siphoned from the flask under an inert atmosphere and the excess 1,3-butadiene was pumped off, leaving a residue which could be purified by sublimation (10^{-3} Torr, 323–333 K) or by recrystallization from heptane. Yields are 50–60% based on the amount of metal deposited in the butadiene matrix. A typical 1 h run yields about 200 mg of pure product {tris(butadiene)tungsten}, which is white, is in the hexagonal system, decomposes at 408 K, is soluble in organic solvents and is stable in air (slight darkening is noticeable after two weeks) [133].

1.4.3 Interaction of metals with arenes and hetarenes

The cryosynthesis procedure is especially effective in the preparation of η^6-arene complexes. Arene π-complexes, the structure of which is adequately described by formula **XXXI**, have been obtained from almost all transition metals and various η^6-aromatic ligand systems, which can clearly donate six π-electrons to a bound metal.

TABLE 1.4
Cryosynthesis of metal complexes with acyclic and cyclic dienes, trienes and tetraenes

Metal	Ligand	Product	T, K	Yield, %	Ref.
(a) Acyclic dienes and trienes					
Fe	$CH_2{=}C{=}CH_2$	$Fe_2(CH_2{=}C{=}CH_2)$	10		46
Fe_2	$CH_2{=}C{=}CH_2$, hv	$Fe(HC{\equiv}C{-}CH_3)$	10		
Mo, W	$CH_2{=}CHCH{=}CH_2$ (L)	ML	77		11, 127, 133
Ni, Fe	Buta-1,3-diene	$M(C_4H_6)$	77, 298 (100 Torr)		45, 49
Cr, Ru	Buta-1,3-diene + L (PF$_3$ CO, Me$_3$CNS)	$M(C_4H_6)_2L$	77		44, 147
Cr	$CH_2{=}CHCH{=}CH_2$ + CO	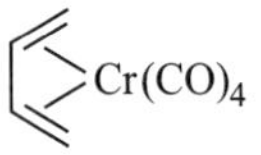	77		11, 127
Co	$CH_2{=}CHCH_2$ $CH{=}CH_2$	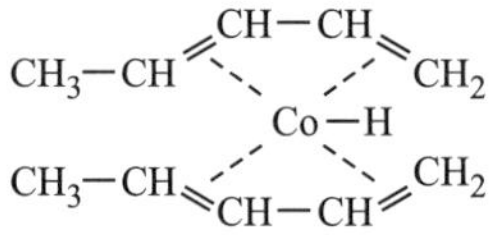	77		11, 16, 127
Cu, Ag, Au	Unconjugated dienes (L)	ML	77		129
Mo	Spiro[2.4]hepta-4, 6-diene	$Mo(C_5H_4CH_2CH_2)_2$	77	50–60	149
(b) Cyclodienes					
Fe, Cr, Mo, W	C_pH	$M(C_p)_2$	77	20–100	16, 147, 150
Fe	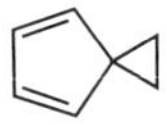	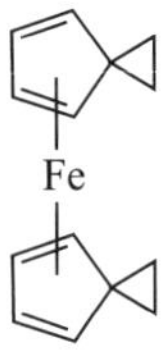	77		151

TABLE 1.4 (Contd.)

Metal	Ligand	Product	T, K	Yield, %	Ref.
Fe	RC_5H_5 $R = Me, Et$	$Fe(RC_5H_4)_2$	77	28–72	16
Mn	$C_pH + CO$	$\{Mn(C_5H_5)(CO)_3\}$ + $\{Mn(\eta^5\text{-}C_5H_5)(\eta^2\text{-}C_{10}H_{12})(CO)_2\}$	77	6–10	152
Fe	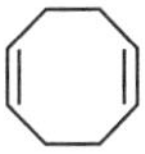	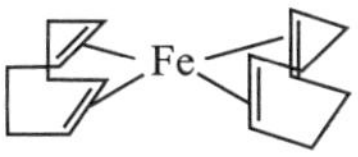	77		74, 153
Ni	Cycloocta-1,5-diene	*Bis*(cycloocta-1,5-diene)nickel	77		148
Co	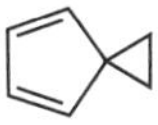 + CO	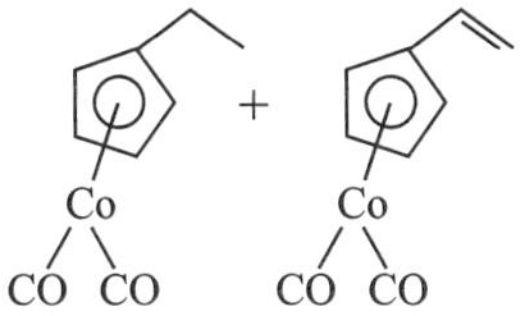	77	1	150
Fe	Dimethylfulvene 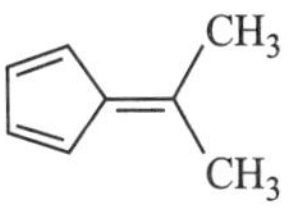	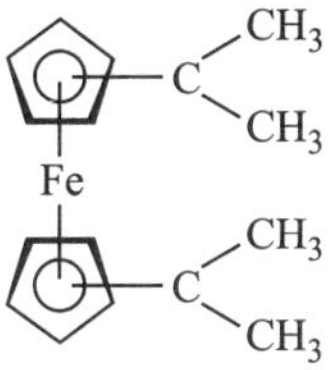 + $[C_5H_4\text{—}C(CH_3)_2H]_2Fe$	77		16
Mn	Cyclohexa-1,3-diene + CO	$\{Mn(C_6H_8)_2(CO)\}$	77	6–10	50
	Cycloocta-1,3-diene + CO	$\{Mn_2(CO)_{10}\}$ + $\{Mn(\eta^3\text{-}C_8H_{13})(CO)_4\}$ + $\{Mn(1\text{–}5\text{-}\eta\text{-}C_8H_{11})(CO)_3\}$			
	Cycloheptatriene + CO	$\{Mn(\eta^4\text{-}C_7H_8)_2(CO)\}$			
Ti, V, Cr, Fe	Cycloheptatriene	ML_2	77	15–46	144
Cr, Co Cr, Fe	Cycloheptatriene + PF_3	$ML(PF_3)_n$			
Ti, Fe, Co, U, Th, Pu	COT	$M(COT)_n$			

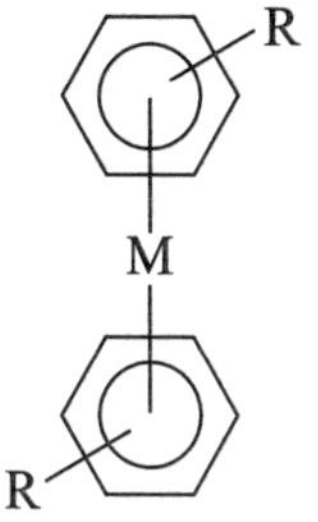

XXXI

The gas-phase synthesis, properties and structures of complexes of type **XXXI** have been reviewed [5,7,12,16] and constitute the subject matter of a monograph [22]. Among early papers, it is necessary to single out the work of Graves [42], who described 38 bis(arene) sandwich complexes of type **XXXI** (R = F, Cl, CF$_3$, Alk, OAlk, NAlk$_2$, COOAlk), obtained by co-condensation of metals and ligands at 77 K and 5 × 10 Torr. The complexes were characterized by ^{13}C NMR spectra and their comparative thermal stability was determined.

^{1}H NMR spectroscopy, along with mass-spectral studies, has been used to characterize chromium complexes **XXXI** (R = F, Cl) containing electronegative substituents in the benzene rings [49]. These compounds were prepared at 77 K and 10^{-4} Torr. Bis(benzene)ruthenium [154], (dicyclopropylbenzene)chromium **XXXII** and its tetraphenylborate derivative **XXXIII** [150] were also synthesized. The reactions take place according to scheme (1.15):

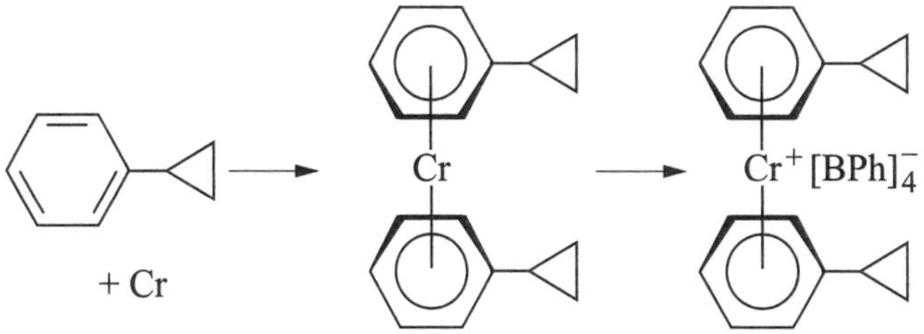

$$(1.15)$$

XXXII   **XXXIII**

In similar conditions, a series of compounds, containing various substituents in the aromatic nuclei (ferrocenyl, yield 1% [155]; CH$_2$Ph 48%, CH$_2$CH$_2$OH 5% and C$_3$H$_7$CO 4% [156]), have been obtained and characterized.

Bis(arene) complexes **XXXIII** having general formula $M(\eta^6\text{-}t\text{-}Bu_3C_6H_3)_2$ [52–54] could be prepared by the reactions of atomic yttrium, titanium, hafnium, niobium, chromium, molybdenum and tungsten with 1,3,5-tri-*tert*-butylbenzene, in spite of apparent steric obstructions. Reactions take place at 77 K and allow the preparation of products with yields of 20% (M = W), 30–50% (M = Cr, Mo) [54], and 40% (M = Ti, Zr, Hf) [52]. In the case of chromium, together with bis(arene) compound **XXXIII** (M = Cr), a three-decker complex with composition $[Cr_2(\eta^6\text{-}t\text{-}Bu_3C_6H_3)_2(\mu\text{-}\eta^6{:}\eta^6\text{-}t\text{-}BuC_6H_3)]$ **XXXIV** [54] is formed (yield 2%).

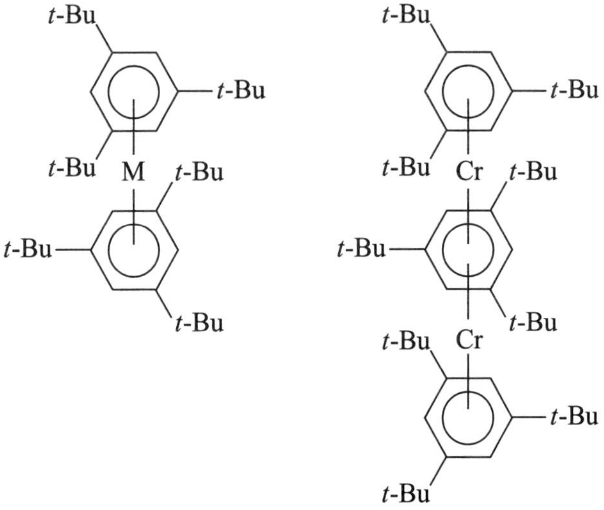

$\quad\quad\quad\quad$**XXXIII**$\quad\quad\quad\quad\quad\quad\quad$**XXXIV**

The reactions of zirconium and hafnium with the same ligand in the presence of carbon monoxide lead to the formation of a stable 18-electron adduct **XXXV** of zirconium and an analogous hafnium complex of low stability. The interaction of the complex **XXXIII** (M = Nb) with $AgBF_4$ in toluene is accompanied by the formation of a tetrafluoroborate complex containing the stable 16-electron cation **XXXVI** [54].

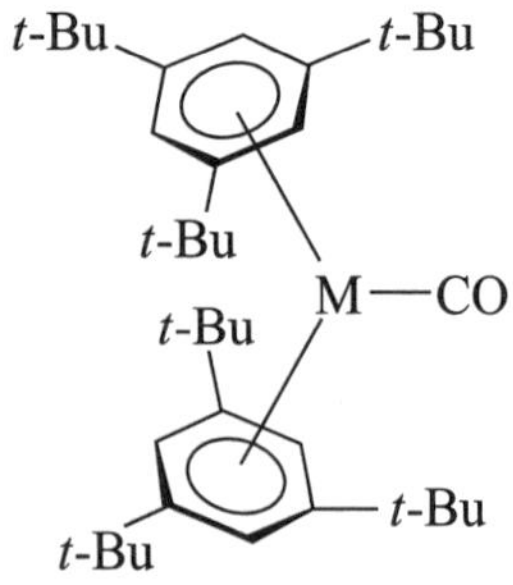

XXXV 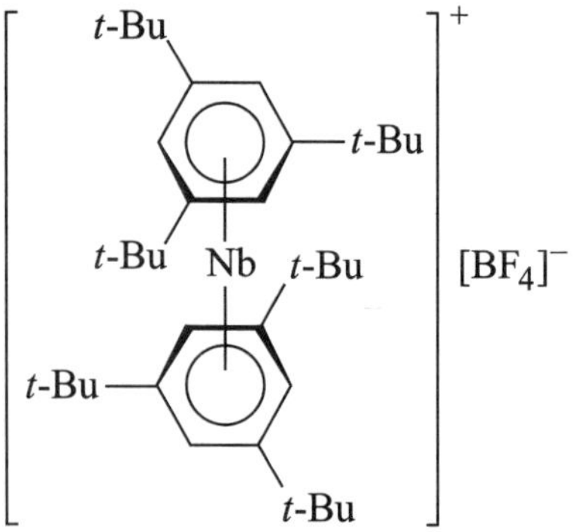   **XXXVI**

Structures of complexes **XXXIII–XXXVI** were proved [54] by mass and ^{1}H NMR spectroscopy. The yttrium complex **XXXI** (M = Y) is unusual, because it is paramagnetic ($\mu_{\text{eff.}} = 1.74\ \mu_B$ at room temperature) and its EPR spectrum in methylcyclohexane shows a well-defined doublet at 77 K: for Y, $g_{\parallel} = 2.085$ (doublet), A = 3.0 mT, $g_{\perp} = 2.005$. A preliminary X-ray diffraction study [53] showed that yttrium and gadolinium complexes contain parallel benzene rings and constitute the first examples of η^6-coordination compounds of these metals.

In spite of the fact that the chromium complex **XXXI** with a dialkylamino substituent (M = Cr, R = NMe$_2$) has been known for more than 20 years [42] and its structure was proved by X-ray diffraction [157], for a long time it was impossible to prepare an analogous complex with aniline. Some years ago [43b], the complex **XXXI** {M = Cr, R = N(SiMe$_3$)$_2$} was obtained by co-condensation of chromium atoms with bis(trimethylsilyl)aminobenzene at 77 K. The synthesized compound, after treatment with tetrabutylammonium fluoride hydrate in THF at 298 K, produced the bis(aniline)chromium complex **XXXI** (M = Cr, R = NH$_2$). The transformation takes place according to scheme (1.16).

$$\text{Cr}_{\text{at.}} + 2\text{C}_6\text{H}_5\text{N}(\text{SiMe}_3)_2 \xrightarrow[\text{co-cond.}]{77\ \text{K}} \textbf{XXXI}\ [\text{R} = \text{N}(\text{SiMe}_3)_2;\ \text{M} = \text{Cr}]$$

$$\Big\downarrow \text{Bu}_4\text{NF}\cdot 3\text{H}_2\text{O}$$

$$\textbf{XXXI}$$
$$\text{M} = \text{Cr}$$
$$[\text{R} = \text{NH}_2]$$

(1.16)

The complex **XXXI** (M = Cr, R = NH$_2$, m.p. 501 K) was characterized by elemental analysis, ^{1}H and ^{13}C NMR and mass spectroscopy; its reduction potential and Bronsted basicity were studied. Upon interaction between bis(π-aniline)chromium and its π-N,N-dimethylaniline analogue with HBF$_4$ in diethyl ether, a protonation takes place on the nitrogen atom.

Bis-π-arene complexes of such metals as titanium have been described [43,158]. Thus, the complex **XXXI** (M = Ti, R = Et) was prepared by the interaction of ethylbenzene with titanium at 77 K in vacuum with ~100% yield [158]. Among the sandwiches of polynuclear arenes, the synthesis and characterization of bis-η^6-naphthalene chromium complexes were reported [159]. In this work a series of coordination compounds having the general formula (η^6-Ar)CrL$_3$, where ArH = C$_{10}$H$_8$, C$_{10}$H$_7$Me, C$_{10}$H$_6$Me$_2$; L = CO, PMe$_3$, P(OMe)$_3$, PF$_3$ (co-ligands), was also synthesized by co-condensation at 193 K of atomic chromium and naphthalene or its methyl-substituted derivatives. The compounds obtained were characterized by ^{1}H NMR spectroscopy. It was emphasized that atomic chromium interacts with 1,4-dimethylnaphthalene regiospecifically, forming mainly (ca 95%) the isomer in which the metal atom is bound to the unsubstituted aromatic nucleus.

Among the most recent investigations, the various metal–arene and–hetarene complexes have been isolated: [(i-PrO)$_2$B-η^6-C$_6$H$_5$]$_2$M (M = Cr,V) bearing boryl substituents [160]; hexakis(η^1-phosphabenzene)-chromium(0), which could be also prepared by ligand substitution at bis(2,4-dimethyl-η^5-pentadienyl)chromium [161a]; [C$_5$(CMe$_3$)$_3$H$_2$P]$_2$M (M = Ho [161b]; M = V,Cr [162]), bis-(η^6-bimesityl)chromium [162]; and numerous π-arene complexes of chromium and other transition metals [163–168], including mononuclear tris(arene)–metal complexes described in a review [169].

A series of studies by Yurieva et al., generalized in a review [56] (see Table 1.5), is devoted to the (bis-arene)chromium complexes of aromatic compounds, containing other groups which are active in relation to the metal atom (e.g. the phosphorus atom in triphenylphosphine substituents, C≡C and C≡N bonds), as well as polynuclear fragments (α, ω-diphenylalkanes, diphenylacetylene). Thus, it is shown that η^6- and η^{12}-coordinated mono- and binuclear chromium complexes are formed as a result of the interaction between atomic chromium and 1,4-diphenylbutane and 1,5-diphenylpentane. This transformation is the first example of the one-step synthesis of the bridge compounds of the dibenzenechromium series.

Using a mixture of benzonitrile with other monosubstituted derivatives of benzene (ArR, R = H, OMe, COMe, CF_3, F, Cl) in the reactions with atomic chromium, a series of complexes of the types $(PhCN)_2Cr$, $(PhCN)Cr\cdot(PhR)$ and $(PhR)_2Cr$ was synthesized with low yields (from 0.6 to 10%). The corresponding tetrasubstituted chromium η^6-complexes were prepared by co-condensation of Cr vapor with disubstituted derivatives of benzene, containing, together with the fluorine atom, Me, MeO and Cl substituents. According to ^{19}F NMR data, two diastereomers are formed as a result of such transformations: the first is a racemic mixture of two enantiomers, the second is a *meso* form [56].

The organometallic compounds of arenes containing metal–carbon σ-bonds could also be prepared by the technique described above. Thus, the complexes $[RuH(\eta\text{-}C_6H_6)(PMe_3)(\sigma\text{-}C_6H_5)]$ and $[OsH(\eta\text{-}C_6H_6)(PMe_3)(\sigma\text{-}C_6H_5)]$ were synthesized by co-condensation of the corresponding metal atoms with C_6H_6 and PMe_3 [170]. Using C_6F_5Br in co-condensation with palladium atoms, an orange–brown powder of the $(C_6F_5PdBr)_x$ (**XXXVII**) is formed, instead of the expected η^6–π-complex with participation of the benzene rings (scheme (1.17)). The product is stable to air and moisture, soluble in organic solvents (where it is found to be dimeric and trimeric), and very reactive with Lewis bases [38].

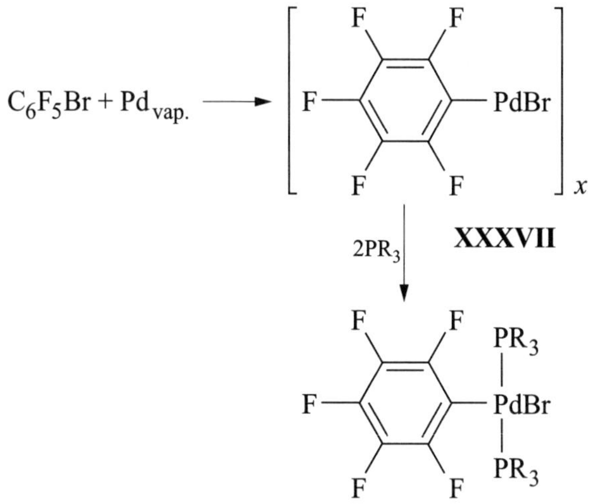

$$ \tag{1.17} $$

The stability, physico-chemical properties and electronic structure of bis-arene complexes, as well as the processes of their syntheses and decomposition, have been studied systematically [5,7,12,16,43,157,171–175]. In particular, the feasibility of using the additive schemes method to calculate the heat of formation ΔH_f of vapor-phase metal bis-arene complexes was examined [175]. The electronic [176,177] and fluorescent [178] spectra of molybdenum complexes were interpreted, and newly measured oxidation–reduction properties of bis-arene derivatives of transition metals [179] were discussed. The activation mechanism of benzene derivatives by lithium atoms is described in recent papers [180,181]. The model reaction is developed using the spectroscopic results and ab initio calculations.

For the example of bis-arene metal complexes, it is possible to present and discuss the mechanism of direct cryosynthesis. It is known [16] that, for the reaction of dibenzenechromium decomposition (1.18), the equilibrium constant increases with temperature in in the range 298–1000 K.

$$(C_6H_6)_2Cr(g) \Leftrightarrow Cr(s) + 2C_6H_6(g) \tag{1.18}$$

The change of magnitude of the Gibbs energy ΔG in the interval 298–800 K and at pressure 10^{-5}–10^{-7} atm was also studied; only at 10^{-7} atm does ΔG change its sign, i.e. equilibrium is moved to dibenzenechromium formation. In order to establish a correct reaction mechanism (the direct reaction could be gas-phase or "gas–solid"), it is possible to calculate the change in ΔG (kcal mol^{-1}; 1 kcal = 4.18 J) for processes (1.19)–(1.26) which may occur in these conditions (chromium atoms are treated in the calculation as an ideal gas):

$$
\begin{aligned}
Cr(s) + 2C_6H_6(g) &\Leftrightarrow (C_6H_6)_2Cr(g) & \Delta G &= 27.2 & (1.19)\\
Cr(s) + 2C_6H_6(g) &\Leftrightarrow (C_6H_6)_2Cr(s) & \Delta G &= 18.9 & (1.20)\\
Cr(g) + 2C_6H_6(g) &\Leftrightarrow (C_6H_6)_2Cr(g) & \Delta G &= -57.4 & (1.21)\\
Cr(g) + 2C_6H_6(g) &\Leftrightarrow (C_6H_6)_2Cr(s) & \Delta G &= -65.5 & (1.22)\\
Cr(g) + 2C_6H_6(liq) &\Leftrightarrow (C_6H_6)_2Cr(s) & \Delta G &= -62 & (1.23)\\
Cr(s) + 2C_6H_6(s) &\Leftrightarrow (C_6H_6)_2Cr(s) & \Delta G &= 16.4 & (1.24)\\
Cr(g) + 2C_6H_6(s) &\Leftrightarrow (C_6H_6)_2Cr(s) & \Delta G &= -70 & (1.25)\\
Cr(g) + 2C_6H_6(g) &\Leftrightarrow (C_6H_6)_2Cr(g) & \Delta G &= -30.3 & (1.26)
\end{aligned}
$$

It is clear that the reactions (1.21)–(1.23), (1.25) and (1.26) are thermodynamically possible. On the basis of the experimental results, it was concluded that equilibrium of reactions (1.19), (1.20) and (1.24) is moved to initial substances. The inactive chromium powder, which is formed as the result of the aggregation of the atomic metal does not react with benzene and remains in the reaction chamber at the end of the process.

Moreover, the thermodynamically permitted reactions (1.21), (1.22) and (1.26) require trimolecular collisions, the possibility of which is almost zero. For gas-phase reactions, the possibilities (z_3) of trimolecular and bimolecular (z_2) collisions are $z_3/z_2 = 10^{-2}p$, where p is the pressure (atm) and $T = 300\,K$. In the conditions of cryosynthesis, p is 10^{-6}–10^{-8} atm, so $z_3 = (10^{-8}$–$10^{-10})z_2$ is extremely low. Taking into account that at this pressure the pathlength of a metal atom is much greater than the dimensions of any type of reaction chamber that may be used, it is evident that the cryosynthesis reaction is a "gas–solid phase" interaction. In this case, the gas is the metal and the solid is benzene. On this basis, the following mechanism of cryosynthesis was proposed [16].

A metal atom, arriving at the surface of the condensed (in)organic substance, evidently already loses its energy in its first collision with condensed-phase molecules (this could be concluded from an evaluation of the kinetic energy of the atom in flight). Local overheating of the condensed phase contributes to diffusion of metal atoms and molecules of the organic phase; as a result, a favorable spatial orientation of the reaction participants in the matrix (for the following interaction) is attained.

When a metal is evaporated rapidly, the possibility of aggregation of its atoms increases. Moreover, insufficient heat exchange and a thick layer of the (in)organic phase on the chamber walls could provoke melting of this phase (reaction (1.23)). The "gas–liquid" reaction is thermodynamically favorable; however, in these conditions the possibility of aggregation of the metal atoms is especially high. Metal in the condensed form does not react. This could explain the reduction of yield of dibenzenechromium and resulting formation of metal powder, although the yield is also determined by reversible decomposition of the product.

In addition to the bis-arene compounds of types **XXXI–XXXIII**, the monoarene, for example **XXXVIII**, and "mixed-ligand" cyclopentadienylarene **XXXIX** complexes were obtained by co-condensation of atomic metals and mixtures of the corresponding ligands [5,16,97].

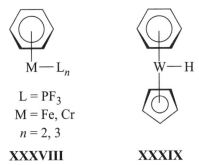

L = PF$_3$
M = Fe, Cr
n = 2, 3

XXXVIII   **XXXIX**

Other "mixed-ligand" compounds of the types **XL–XLIII**, which have been synthesized by conventional chemical methods [38], theoretically could also be obtained by interaction of metal atoms with a mixture of the corresponding ligands (see [56] and Table 1.5).

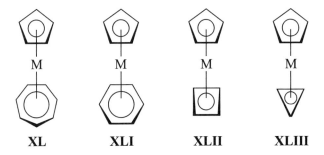

XL   **XLI**   **XLII**   **XLIII**

Unusual iron(IV) complexes **XLIV** with the composition (η^6-benzene)Fe(H)$_2$(SiCl$_3$)$_2$, (η^6-toluene)Fe(H)$_2$(SiCl$_3$)$_2$ and (η^6-p-xylene)Fe(H)$_2$(SiCl$_3$)$_2$ were obtained from "solvated iron atoms" in the corresponding solvent (benzene, toluene or p-xylene); these were prepared by co-deposition of solvent with iron vapor, and HSiCl$_3$ [182]. The novel π-arene products were characterized by X-ray diffraction data and ^{1}H NMR spectroscopy. These results show that there is no chemical interaction between Si and H or between H and H atoms in these compounds [182].

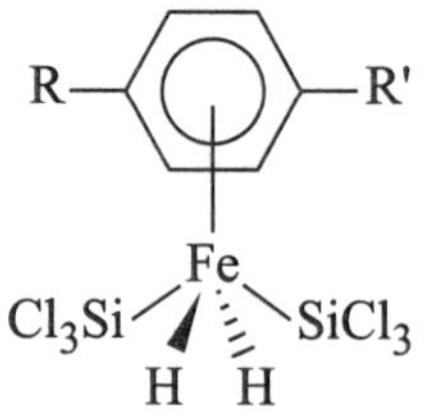

R—⟨○⟩—R'

Fe
Cl₃Si⟋ ⋮ ⟍SiCl₃
   H   H

R = R' = H
R = H, R' = Me
R = R' = Me

XLIV

In addition to the noted hexakis(η^1-phosphabenzene)chromium(0) and bis(2,4,6-tri-tert-butylphosphine)holmium(0) [161], complex compounds of heteroaromatic systems, obtained in cryosynthesis conditions, are represented by π-complexes of pyridine and its C-methyl-substituted derivatives. In fact, the possibility of participation of the pyridine π-system as a ligand was first shown when the π-complex **XLV** was prepared by co-condensation of atomic chromium with pyridine in the presence of PF$_3$ as a co-ligand [7]. The low-temperature reaction between atomic Cr and 2,6-dimethylpyridine led to the first sandwich π-complex of an azine type (although with a low yield—2%): bis(2,6-dimethylpyridine)chromium **XLVI** (R = Me) [57]. Its structure was proved by X-ray diffraction [57,183]. Similar compounds were isolated using Ti, Mo or V vapor [184]. Wucherer and Muetterties [184] noted that attempted syntheses of bis(η^6-2,6-dimethylpyridine)metal complexes of other metals were unsuccessful, probably as a result of reaction pathways involving nitrogen lone pair–metal σ or CH–metal interactions. Using trimethylsilyl derivatives of pyridine in the same synthesis, it is also possible to prepare a parent of the bis(η-azine) metal complexes: bis(η-pyridine) **XLVI** (R = H) [185]. We emphasize that this proved the impossibility of obtaining any of the bis(η^6-pyridine)metal complexes by the conventional synthetic procedures [186,187].

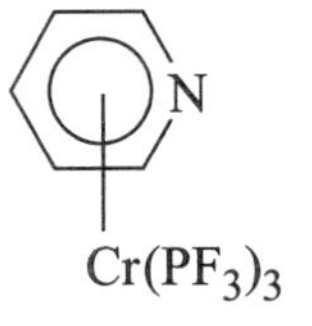

XLV   XLVI

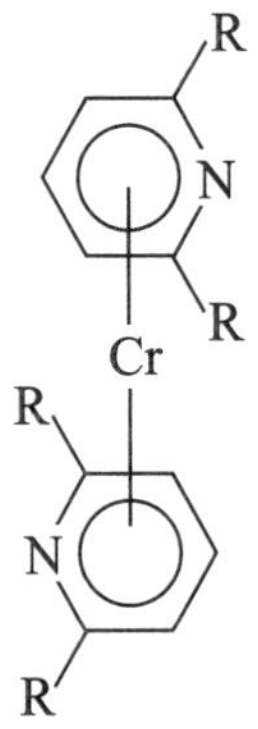

At the same time, it is necessary to take into account the fact that the interaction of azines with atomic metals under cryosynthesis conditions could also lead to N–σ-complexes. Precisely this type of coordination compound was obtained by co-condensation at 77 K of chromium vapor with α, α'-bipy **XLVII** (bipy = bipyridyl) [58].

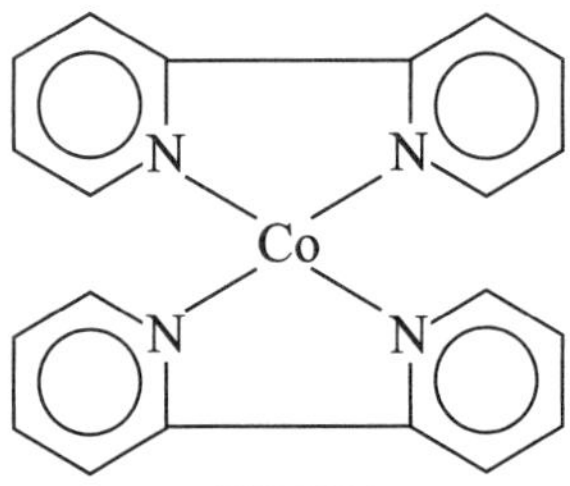

XLVII

The properties of this compound were studied by IR, UV and EPR spectroscopy, as well as by determination of its magnetic properties [58]. Its μ_{eff}, determined by Faraday's method, was 2.02 μ_{B}, which is probably related to the electronic configuration of **XLVII**, containing one unpaired electron. This complex easily undergoes oxidation chemically (with Br_2, tetracyanoethylene or tetracyanoquinodimethane) and electrochemically [58].

TABLE 1.5
Cryosynthesis of metal complexes with arenes and hetarenes

Metal	Ligand	Product	T, K	Yield, %	Ref.
Cr, Nb, V, Cr, W, Ti etc.	PhR, $R = Alk, CF_3$, OR, Hal, NR_2 COOR, H	$M(PhR)_2$	77	2–60	11, 16, 56, 106, 147, 151, 191, 194
Fe, Cr	C_6H_6, PF_3	$Fe(C_6H_6)(PF_3)_2$	77		188
Ni	C_6F_6	$Ni(C_6F_6)$	77		191, 195
Cr	$PhCH_3$ PhC_2H_5	$Cr(PhCH_3)_2$ $Cr(PhC_2H_5)_2$	77 77	60 52	147
K	Arene, THF, MX_n ($M = Ti, V, Cr, Mo$ $X = Hal$)	$M(arene)_2$	173		6
K	$MoCl_5 + PhCH_3$ $VCl_3 \cdot 3THF +$ toluene	$Mo(PhCH_3)_2$ $V(PhCH_3)_2$	173	40 30	192, 196
	$TiCl_3 \cdot 3THF +$ toluene	$Ti(PhCH_3)_2$		20	
	$TiCl_3 \cdot 3THF +$ mesitylene	$Ti(C_6H_3Me_3)_2$		15	
Cr	Cumene	$Cu(cumene)_2$	77	19	188
Cr	C_6H_6, C_6F_6	$Cr(C_6H_6)(C_6F_6)$	77		106
Cr	$o\text{-}ClC_6H_4CF_3$	$Cr(o\text{-}ClC_6H_4CF_3)_2$	77	33	191, 192

TABLE 1.5 (Contd.)

Metal	Ligand	Product	T, K	Yield, %	Ref.
Cr	PhCN	Cr(PhCN)$_2$	77		56
Cr	PhCN + PhX (X = H, CH$_3$, OCH$_3$, Cl, F, CF$_3$, CN)	[Cr complex structures: Cr(C$_6$H$_5$CN)(C$_6$H$_4$X) + Cr(C$_6$H$_4$CN)$_2$ + Cr(C$_6$H$_4$X)$_2$]	77		56
Cr	p-XC$_6$H$_4$F (L) X = CH$_3$, CO$_2$CH$_3$, Cl	CrL$_2$	77		56
Pt, Pd	C$_6$F$_5$X, PEt$_3$ X = Cl, Br	(PEt$_3$)$_2$MX$_2$+(C$_6$F$_5$)$_2$(PEt$_3$)$_2$ (PEt$_3$)$_2$M(C$_6$F$_5$)X	77	3–66	197, 198
Pt	C$_6$H$_5$Br, PEt$_3$	(C$_6$H$_5$)$_2$(PEt$_3$)$_2$Pt	77	3	198
Ni, Co	C$_6$F$_5$Br, C$_6$H$_6$	M(C$_6$H$_6$)(C$_6$F$_5$)$_2$	77		106
Cr	PhCH$_2$OCH$_2$Ph	Cr(PhCH$_2$OCH$_2$Ph)	77		56, 74
Cr	[anthracene/triphenylene ligand structure]	[Cr bis(arene) complex structure]	77		74

TABLE 1.5 (Contd.)

Metal	Ligand	Product	T, K	Yield,%	Ref.
Cr	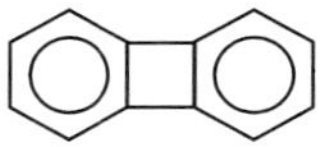	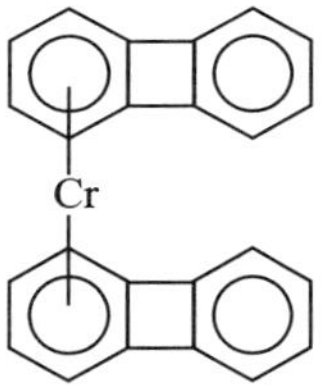	77		74
Cr		Bis(η^6-naphthalene)chromium(0)	193	43	159, 199
K	CrCl$_3$(THF)$_3$ + naphthalene	Cr(C$_{10}$H$_8$)$_2$	163	36	196
	VCl$_3$(THF)$_2$ + 1-methylnaphthalene	V(C$_{10}$H$_7$Me)$_2$		25	
	CrCl$_3$(THF)$_2$ + 1-methylnaphthalene	Cr(C$_{10}$H$_7$Me)$_2$		40	
	MoCl$_5$ + 1-methylnaphthalene	MO(C$_{10}$H$_7$Me)$_2$		42	

TABLE 1.5 (Contd.)

Metal	Ligand	Product	T, K	Yield, %	Ref.
Cr	PPh$_3$	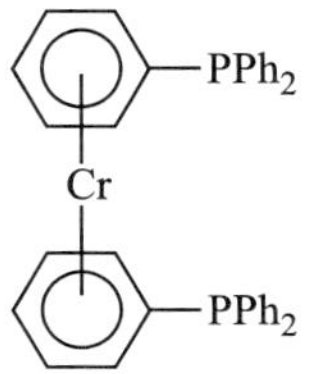	77		56
Cr	PhC≡CH	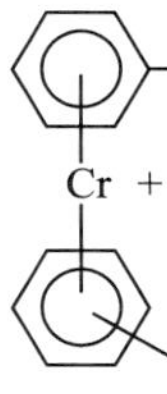	77		56
Cr	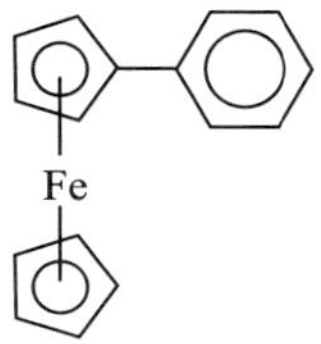	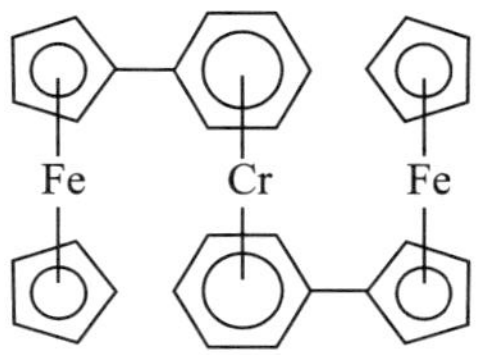	77		56

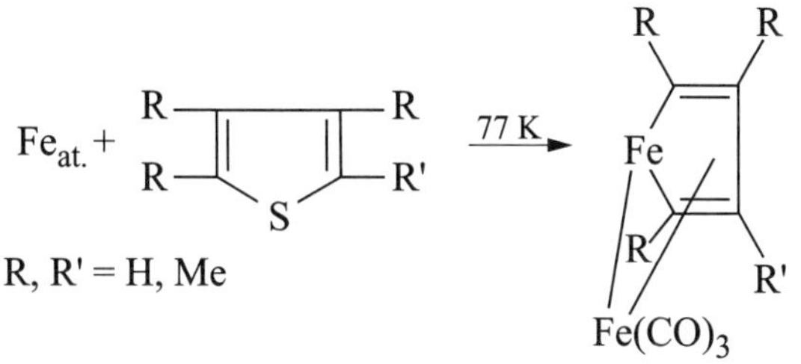

$$R, R' = H, Me$$

XLVIII   XLIX

$$(1.27)$$

Although a review [188] indicates that thiophene does not react with elemental metals, the transformation (1.27) is known [22]; it is accompanied by desulfuration of the heterocycle **XLVIII** and formation of bimetallic complex **XLIX**. This fact testifies to fundamental rebuilding in the process of complex formation with participation of the thiophene system; it should be taken into consideration in further research on gas-phase synthesis with sulfur-containing heterocycles as ligands. According to [188–190], these heteroaromatic systems have a real π-donor activity.

Summarized data on vapor synthesis of metal– arene and –hetarene complexes are presented in Table 1.5.

Experimental procedures

V + trifluorobenzene

Vanadium was vaporized from a tungsten boat. Simultaneously trifluorobenzene vapor was co-deposited with the vanadium vapor on the liquid-nitrogen cooled walls of the glass reactor. Co-deposition continued for 10 min (35.4 mg, 0.70 mg-atom of vanadium; 10 ml of trifluorobenzene). After completion of the reaction the cold matrix was red–black. It was allowed to warm slowly to room temperature and then pressurized with argon. With argon flushing, the solution was removed by syringe and transferred to a sublimer. The excess ligand was slowly pumped off, and then the complex was sublimed to yield 51.2 mg (22%) of bis(trifluorobenzene)vanadium(0) [192].

Other metal–fluorocarbon compounds that have been prepared and characterized [12] are presented in Tables 1.6 and 1.7. Some interactions between Ni and C_6F_6 are presented in Fig. 1.6.

Cr + naphthalene

Chromium (0.98 g, 18.8 mmol) was evaporated over a period of 2.5 h from an alumina crucible (1 cm^3) with imbedded molybdenum wire, resistance-heated. The

TABLE 1.6
Synthesis of *bis*(arene)vanadium(0) and *bis*(arene)chromium(0) employing the metal-atom technique [12]

Compound	Yield,* %	Properties
$(C_6H_5F)_2V$	13	Red, air-sensitive, sublimes 50–70°C/10^{-3} Torr
$(C_6H_5Cl)_2V$	7	Orange–red, air-sensitive, sublimes 50–60°C/10^{-3} Torr
$(p\text{-}F_2C_6H_4)_2V$	1	Red, air-sensitive, sublimes 25°C/10^{-3} Torr, odor of organic sulfide
$(C_6H_5CF_3)_2V$	22	Orange, moderately air-sensitive, sublimes 50–60°C/10^{-3} Torr (80–90°C decomp.)
$(C_6H_5CF_3)_2Cr$	26	Yellow, m.p. 91–91.5°C, decomp. 170°C, slightly air-sensitive
$[m\text{-}(CF_3)_2C_6H_4]_2Cr$	17	Yellow–green, m.p. 91.5–92.5°C, decomp. 200°C, sublimes rapidly at 100°C/10^{-2} Torr, very air-stable
$[p\text{-}(CF_3)_2C_6H_4]_2Cr$	38	Amber, m.p. 150–152°C, decomp. 266°C, sublimes rapidly at 100°C/10^{-2} Torr, very air-stable
$[o\text{-}(CF_3)_2C_6H_4]_2Cr$	33	Yellow–green, m.p. 81–82.5°C, decomp. 165°C, sublimes rapidly at 100°C/10^{-2} Torr, very air-stable

*Based on metal vaporized.

TABLE 1.7
Decomposition temperatures (°C of metal compounds with C_6H_6 and C_6F_6 [195]

	Ti	V	Cr	Fe	Co	Ni
C_6F_6	−50	100	40	−40	10	70
C_6H_6	−50	> 300	> 300	−50	—	—

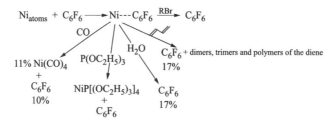

Fig. 1.6. Reactions of Ni atoms with C_6F_6 [195].

chromium vapor was condensed into a cooled (193 K) solution of naphthalene (10 g, 78 mmol) in dry diglyme (150 cm^3) in a rotating flask (2 dm^3). The pressure in the reactor was kept below 10^{-2} Torr during the reaction. On evaporation of chromium, the solution quickly turned an intense reddish brown color. After removal of the solvent in vacuo, the excess of naphthalene was sublimed onto a liquid-dinitrogen cooled cold-finger (1 h, 313 K, 10^{-3} Torr). The black residue was extracted with toluene (3 × 100 cm^3), filtered, diluted with pentane (200 cm^3), and cooled to 195 K to give black crystals of bis(η^6-naphthalene)chromium(0) on standing for 48 h. The crystals were separated, washed with cold pentane and dried in vacuo (2 h, 10^{-3} Torr). Yield was 2.49 g, 43% based on chromium [159].

K + [TiCl₃(THF)₃]

[TiCl$_3$(THF)$_3$] (2.5 g, 6.75 mmol) was dissolved in THF (180 cm^3) and toluene (40 cm^3) and the pale blue solution was treated with an excess (2.0 g, 50 mmol) of potassium vapor condensed into the solution at 163 K in 50 min. As K atoms condensed into it, the solution turned pale green, then a cloudy pale brown, and finally a dark, yellowish green solution containing suspended metal was obtained. The flask was rotated for 15 min at 173 K after the metal had ceased evaporating. The solvent was then pumped off and the residual black solid taken up in THF (150 cm^3). This solution/suspension was filtered, first through glass wool and then onto an 8-cm column of alumina (degassed for 12 h at 373 K under high vacuum before use) above a frit. The solution changed in contact with the alumina and a pink solution passed through. The THF was removed under vacuum and the red residue recrystallized from pentane to give [Ti(C$_6$H$_5$Me)$_2$] (300 mg, 1.28 mmol, 19% yield) [196b].

Fe + C₆H₆ + HSiCl₃

Iron vapor (1.2 g) was co-deposited with benzene (about 90 g) at 77 K. HSiCl$_3$ (about 10 cm^3) was distilled in and the matrix warmed to room temperature over 5–6 h. The solution was stirred overnight. The reaction mixture was filtered through

Celite under argon. The volatiles were removed under reduced pressure, the resultant yellow solid was dissolved in benzene, and the solution was layered with hexanes at room temperature. Yellow prismatic crystals of the $(\eta^6$-benzene)Fe(H)$_2$(SiCl$_3$)$_2$ **XLIV** were formed on the wall of the Schlenk tube within two weeks (1–2% yield based on Fe vaporized) [182].

Co + α, α'-bipy

During a 1.5-h period 0.8288 g of Co metal was deposited at 77 K with 200 cm^3 of dry, degassed toluene, yielding a yellow–brown matrix. Then a solution of 3.02 g of α, α'-bipy in 25 cm^3 of toluene was syringed in under a flow of N$_2$. The contents of the reactor were allowed to warm with stirring. During this time, the matrix formed a yellow–brown solution on meltdown (cobalt–toluene). After addition of the α, α'-bipy-toluene, the color changed to black. The green–black solution was syringed out and filtered through medium-porosity fritted glass, and toluene was removed under vacuum to a volume of 20 cm^3. Crystals were obtained by cooling at 273 K. The crystals were isolated by decanting the mother liquor and were then dried under vacuum; yield was 2.228 g (62% based on α, α'-bipy added as limiting reagent) [58].

Cr + 2,6-dimethylpyridine

Bis(2,6-dimethylpyridine)chromium was prepared by co-condensing chromium atoms with the ligand at 77 K, the latter substance being present in excess. The frozen matrix of ligand containing metal atoms underwent the typical color change observed in other syntheses of this type (colorless → dark) as it warmed to room temperature. The final reaction mixture was a red–brown solution in excess ligand containing dispersed unreacted metal. Excess ligand was removed in high vacuum, and the dark residue was sublimed to give a red–brown substance, as the sole product, in 2% yield (based upon chromium) which proved to be bis(2,6-dimethylpyridine)chromium (m.p. 79–80°C) [57].

1.4.4 Interaction of metals with alkynes

Interaction of atomic metals with acetylene and its substituted derivatives under cryosynthesis conditions leads mainly to formation of complexes with the general formula ML$_n$, where L is C≡CR, and products of tri-, tetra- and polymerization of alkynes [6,14,16,22,33,45,46,152,200]. It was considered earlier [22] that, as a result of these transformations, benzene and its substituted derivatives are the main products, and that only some reactions, for example (1.28) and (1.29), lead to adducts with metals [6,22].

$$M + F_3CC \equiv CCF_3 \xrightarrow[\text{co-cond.}]{77\,K} ML_n \text{ (unstable)} \xrightarrow[\text{heating to 298 K}]{CO,\,77\,K} M(CO)_2L$$

$$\downarrow \text{r.t.}$$

$$M_4(CO)_4L_3 \qquad (1.28)$$

$$Fe + Me_3SiC \equiv CSiMe_3 \xrightarrow[\text{co-cond.}]{77\,K} Fe(Me_3SiC \equiv CSiMe_3) \qquad (1.29)$$

However, a large variety of complexes of alkynes ML_n are presented in a review [14]. They could be formed from almost all the metals of the Periodic Table. Thus, alkali metals form complexes with acetylene in argon matrices [201]. The adducts of aluminum, gallium, copper, and nickel [200] have also been described. An ESR spectroscopic study of the reaction of ground-state Al atoms with acetylene in cyclohexane and adamantane at 77 K demonstrated that $trans$-η^1-aluminovinyl AlCHCH is the initial paramagnetic product, which disappears on warming to 100 K. A second aluminum species is formed in adamantane; it has magnetic parameters which indicate that it is not the π-complex η^2-Al(CHCH) or aluminovinylidene, AlCHCH$_2$, but it may be a hydrogen-bridged isomer. Ga atoms with acetylene give the species Ga(CHCH), with two magnetically equivalent H and C nuclei; it is either an η^2-, π- or σ-complex [202].

The compounds $M(\pi$-$C_2H_2)_n$, where n = 1 or 2, M = Ni [200], Fe [45] or Cu [33,200], are formed by co-condensation of these transition metals with acetylene at 10 K. It is interesting that the complex of iron with monomethylacetylene L, having the same composition, has also been obtained by reaction of allene with iron atoms at 10 K (argon, irradiation by visible light) [46]. The reaction takes place according to scheme (1.30).

$$Fe + \begin{array}{c} H \\ \diagdown \\ H \diagup \end{array} C = C = C \begin{array}{c} H \\ \diagup \\ \diagdown H \end{array} \xrightarrow[\text{visible light}]{} \underset{\textbf{L}}{Fe(HC \equiv C - CH_3)} \qquad (1.30)$$

Under UV irradiation L yields two σ-organometallic complexes: H—Fe—C≡C—CH$_3$ and H$_3$C—Fe—C≡C—H [46]. The formation of σ-complexes under cryosynthetic conditions is also observed for other metals [14].

As shown above (Section 1.4.2) mixed-ligand complexes, for example $C_5H_5Mn(HC{\equiv}CH)_5$ [48], could be formed by co-condensation of acetylene, cyclopentadiene and atomic metals under the conditions of cryosynthesis.

A series of metal–alkyne complexes of the type $L_2M(RC_2R')$, where $L = Ph_3P$, $M = Ni$, Pd, Pt, $R = R' = CF_3$; $L = Ph_3P$, $M = Ni$, Pt, $R = R' = Ph$, and $R = Ph$, $R' = Me$; $L = Ph_3P$, $M = Pd$, Pt, $R = R' = COMe$; and $R = R' = CF_3$, $M = Pd$, Pt, $L = n\text{-}Bu_3P$, Me_2PhP have been described [203]. The IR spectra indicate strong bonding of the alkyne molecule to the metal.

Various types of di-, oligo- and polymerization which take place with participation of atomic metals and alkynes in gas-phase reactions have been summarized in the reviews [14,16]. As an example, atomic nickel in its reactions with $RC{\equiv}CR'$ ($R,R' = H$, Me, Et, Bu, Ph) forms Ni–alkyne polymers, substituted with benzene and COT [14]. A closely similar situation is also observed in the case of co-condensation between ethynes and atomic iron [14].

The bonding in alkyne complexes of various metals is discussed in detail in [14]. Among recently published work, [204], in which the bonding of acetylene to copper atom, dimer, and trimer is discussed, is noteworthy. It was shown that the $Cu–C_2H_2$ complex has a C8 structure and a bonding energy (BE) of 10 kcal mol^{-1} (41.8 kJ mol^{-1}). Three isomers of $Cu_2C_2H_2$ have similar total energies: a C_{2v} end-bonded structure with a BE of 18 kcal mol^{-1}, and two 1,2-dicuproethylene isomers—a *cis* form with a BE of 12 kcal mol^{-1} and a *trans* form with a BE of 15 kcal mol^{-1}. Two stable isomers of $Cu_3C_2H_2$ were also found [204].

Summarized data on vapor synthesis with participation of alkynes are reported in Table 1.8.

Experimental procedures

Cu + acetylene

Monoatomic copper vapor was generated by directly heating a tungsten-rod assembly around which copper wire was wrapped. The rate of metal atom deposition (10–12 K) was monitored continuously by means of a quartz crystal microbalance assembly in situ. $Cu(C_2H_2)$ and $Cu(C_2H_2)_2$ were detected by spectroscopic methods [200].

TABLE 1.8
Cryosynthesis of metal complexes with alkynes

Metal	Ligand	Product	T, K	Yield, %	Ref.
Li, Na, K, Cs	C_2H_2	$M^+\{HC_2H\}^-$	77		14
Al	C_2H_2	3 possible isomers of the Al–C_2H_2 adducts The vinylidene adduct	77		14

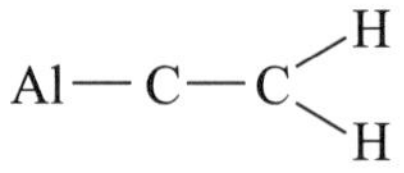

$$Al-C-C\overset{\textstyle H}{\underset{\textstyle H}{<}}$$

The *cis*-vinyl adduct

$$\overset{\textstyle H}{\underset{\textstyle Al}{>}}C-C\overset{\textstyle H}{}$$

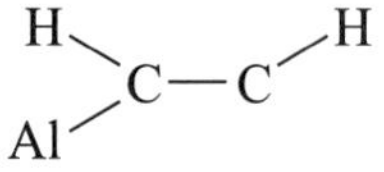

The *trans*-vinyl adduct

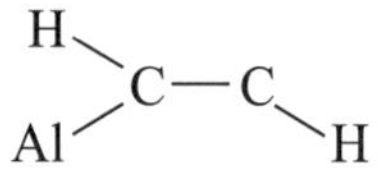

$$\overset{\textstyle H}{\underset{\textstyle Al}{>}}C-C\overset{}{\underset{\textstyle H}{}}$$

Metal	Ligand	Product	T, K	Yield, %	Ref.
Ni, Cu	C_2H_2	$M(C_2H_2)_n$, $n = 1, 2$	10–12		200
Au	C_2H_2	$Au(C{=}CH_2)$	77		14
Mn	$C_2H_2 + CpH$	$(Cp)Mn(C_2H_2)_2$	77		48
Er	C_2H_2	$\{ErC_9H_{15}\}_n$, $n = 2, 10$	77		14
Ge	C_2H_2	$(C_{2}H_{2.7}Ge_{0.72})$ polymer	77	66	205
Sn	C_2H_2	$(C_{2}H_{2.6}Ge_{0.70})$ polymer		29	
Co	C_pH+ but-2-yne + penta borane(9)	$\{2,3\text{-}(CH_3)_2\text{-}1,2,3\text{-}(\eta^5\text{-}C_5H_5)CoC_2B_4H_4\}$ $\{2,3\text{-}(CH_3)_2\text{-}1,7,2,3\text{-}(\eta^5\text{-}C_5H_5)_2Co_2C_2B_3H_3\}$ $\{2,3\text{-}(CH_3)_2\text{-}1,7,2,5\text{-}(\eta^5\text{-}C_5H_5)_2Co_2C_2B_5H_5$	77		14
Fe	$CH_3C{\equiv}CCH_3$	$Fe(CH_3C{\equiv}CCH_3)_5$	77		14
Fe, Ni	$C_2H_5C{\equiv}CCH_3$	$Fe(C_2H_5C{\equiv}CCH_3)_5$ $Ni/C_2H_5C{\equiv}CCH_3$ polymer			
Ni, Pd	$CF_3C{\equiv}CCF_3$ +CO	$M(CO)_2CF_3C{\equiv}CCF_3$	77		6
Sm	Hex-3-yne	SmC_6H_{10}	77		14
Yb		YC_6H_{10} probably			
Fe	$PhC{\equiv}CCH_3$	$Fe(PhC{\equiv}CCH_3)_5$	77		14

Ge + acetylene

0.642 g (8.85 mmol) of germanium metal was co-condensed with 0.146 mol of acetylene gas over aproximately 2.5 h. Then the reactor was isolated from the dynamic vacuum and allowed to warm to room temperature under a static vacuum. The unreacted acetylene was recondensed back into the storage bulb and degassed, and the amount of acetylene in the bulb was measured (80.137 mol of acetylene unreacted after reaction; 8.80 mol of acetylene consumed). The reactor was filled with dry nitrogen, and 150 cm^3 of dry toluene was added by a syringe. The insoluble polymeric product, as a slurry in toluene, was siphoned out of the reactor, under nitrogen, into a Schlenk tube, the tube was capped, and the toluene was removed from the polymer in vacuo. The dry product was a yellow–brown powder, in a yield of 0.646 g based on germanium vaporized and with an approximate formula of $(C_2H_3Ge_{0.7})_x$ (66%) [205].

1.4.5 Interaction of metals with other ligands

Complexes obtained from atomic metals and oxygen-containing ligands are represented by comparatively few examples. Phenols and aromatic alcohols, ethers, ketones and acids behave as normal η^6-ligands, forming bis(η^6-arene) metal-substituted derivatives of type **XXXI** (R = CH$_2$CH$_2$OH [156], OAlk [13,16], COAlk [156], COOAlk [16]; Alk = Me, Pr). The alcohols form metal alcoholates through cryosyntheses [16]. The stable complexes [59], forming colloid systems in non-aqueous solutions with particle size ~8 nm, are obtained by interaction between acetone and atomic palladium. It is proposed [59] that in this case reactions (1.31) take place.

$$n\,Pd + x\,H_3C - \underset{\underset{O}{\|}}{C} - CH_3 \xrightarrow{77\,K} Pd(CH_3COCH_3)_x$$
$$\downarrow$$
$$Pd_n(CH_3COCH_3)_y \tag{1.31}$$

However, the structures of the products have not been studied in detail. Oxidative addition (1.32) takes place when palladium interacts with trifluoroacetic anhydride in the presence of triphenylphosphine [16].

$$Pd + \begin{matrix} CF_3-C\diagdown^O_O \\ \\ CF_3-C\diagdown^O_O \end{matrix} + 2PPh_3 \longrightarrow \left[cis\text{-}(PPh_3)_2Pd \begin{matrix} CF_3-C\diagdown^O_O \\ \\ CF_3-C\diagdown^O_O \end{matrix} \right]_2 \qquad (1.32)$$

$$\textbf{LI}$$

It is known [12,13] that the interaction of acyl halides with atomic metals also leads to insertion of the metal into the C–Hal bond. An example of such a reaction is the transformation (1.33).

$$Pd_{at.} + nRCOCl \rightarrow nRCOPdCl \, (R = CF_3, C_3F_7) \qquad (1.33)$$

As a result of the co-deposition of Co vapor with C_6F_5Br, the π-complex $Co(C_6F_5)_2$ is formed, workup of which in toluene yields [η^6-$C_6H_5CH_3(C_6F_5)_2Co$] [193]. An X-ray structure of this compound reveals two σ-bonded fluorophenyl rings with a Co–C bond distance of 1.931 Å; the Co–C(toluene) π-bond distance is 2.141 Å. However, in the case of the interaction between Pd atoms and the same ligand, the σ-organometallic complex $(C_6F_5PdBr)_x$ is obtained [38,197] (see Section 1.4.3).

Metal insertions have also been performed in the preparation of some RMHal, ArMHal and RCOMHal complexes with other metals (M = Al, Ni, Pt, Co, Fe, Mn etc.), in particular from highly active metal slurries obtained by metal atom–solvent co-condensations [74,206]. Synthesis, isolation and reactions of coordinatively unsaturated organometallic compounds containing both σ- and π-bonds have been described [207].

Various perfluoroalkylzinc halides with composition RZnI have been prepared from the zinc atoms at liquid-nitrogen temperatures [74,208]. These compounds are unstable and decompose rapidly upon warming (77 K). It is noted that only perfluoroalkyl iodides react with zinc vapor [74].

The data on reactions of metal vapors with (in)organic **radicals** leading to the metal alkyls and other σ-bonded metal compounds have been reported [209–211]. Thus, the radicals ·SiF$_3$, obtained from a glow discharge of hexafluorodisilane, react at low temperatures with metal atoms giving $M(SiF_3)_n$ (M = Hg, Te, n = 2; M = Bi, n = 3) in moderate to high yields [210]. With Pd and Ni vapors in similar conditions the radicals ·CF$_3$ give unstable unsaturated $M(CF_3)_2$ which, when stabilized by trapping with PMe$_3$ at low temperatures, give trans-$(CF_3)_2Pd(PMe_3)_2$ and $(CF_3)_2Ni(PMe_3)_3$ [211].

Data on vapor syntheses with organohalide derivatives and acyl anhydrides are summarized in Table 1.9.

Experimental procedures

Ni + allyl bromide (allyl chloride)

About 1.5 g of nickel was vaporized at 1823 K over 30 min from a resistively heated alumina-coated molybdenum wire spiral inside an evacuated 200-mm diameter glass vessel which was partly immersed in liquid nitrogen. About 20 g of the allyl halide was simultaneously vaporized into the vessel and condensed with the nickel vapor on the cold walls. During this co-condensation, the pressure in the vessel was below 2×10^{-4} Torr so that few gas-phase intermolecular collisions occurred.

When the nickel had vaporized, the vessel was warmed to room temperature and the excess allyl halide was pumped off. The vessel was then warmed to 343 K and the volatile π-allylnickel halide pumped out into an adjoining cooled trap. The isolated yield of the π-allylnickel bromide was 2.7 g (60% yield based on nickel vapor). In the case of π-allylnickel chloride a 75% yield was obtained, but the product was contaminated by 1–2% of a complex mixture of C_{10}–C_{15} hydrocarbons [215].

Sn + CH₃I (use of tin slurries prepared from tin vapor)

Tin slurries were prepared by the co-deposition of tin vapor ($\approx$0.5 g) with about 50 cm³ of solvent in about 1 h. The colors of the matrices varied depending on whether the solvent was toluene (yellow), dioxane (brown–black), THF (black) or hexane (black). On warming the solution, finely divided black slurries were formed and were transferred under fast N_2 flushing to a 100-cm³ one-necked round-bottom flask equipped with a condenser, magnetic spin bar and inert gas inlet. Methyl iodide (50 mmol) was added via syringe and the reaction mixture heated to reflux for 21 h. All of the metal was consumed, and a yellow precipitate was formed (SnI_4). The reaction mixture was filtered to remove SnI_4, the toluene was evaporated under vacuum, the resultant viscous liquid $\{(CH_3)_x SnI_{4-x}\}$ was dissolved in 25 cm³ of CCl_4, and a weighed amount of CH_3CN was added to the solution as an NMR standard. CH_3SnI_3, $(CH_3)_2SnI_2$ and $(CH_3)_3SnI$ were formed with an overall yield of 51% of organotin compounds based on the amount of tin vaporized. A 14% yield of SnI_4 was also isolated [206].

Active Mg preparation

In a typical active Mg preparation, 3.5 g Mg was vaporized in 1 h from an aluminum oxide crucible in a tungsten wire heater. This apparatus was contained

TABLE 1.9

Cryosynthesis of metal complexes with organohalide derivatives and acyl anhydrides

Metal	Ligand	Product	T, K	Yield, %	Ref.
Pd	$PhCH_2Cl$	$PhCH_2PdCl$	77		13
Pd	$n\text{-}C_3F_7COCl$ CF_3COCl C_6F_5Br	$C_3F_7COPdCl$ $CF_3COPdCl$ C_6F_5PdBr	77		191
Ca Zn	C_6F_6 CF_3I	C_6F_5CaF CF_3ZnI	77		191
Fe	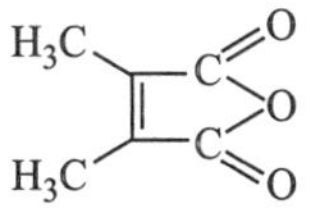	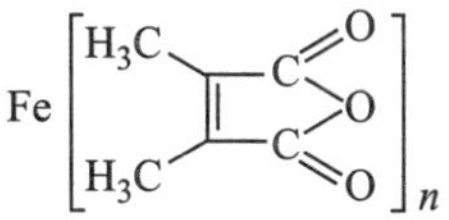	77	10–15	16
Fe	PhBr	$(PhPdBr)_n$	77		106
Ag	$(CF_3)_2CFI$	$(CF_3)_2CFAg$	77		151
Ag	$(CF_3)_2CFI,$ $c\text{-}C_4H_8,$ CH_3CN	$(CF_3)_2CFAg \cdot CH_3CN$	77	5	212
	CF_3I C_3F_7I	CF_3AgI $CF_3CF_2CF_2Ag$	195 195	10	
$Mg(^1S)$	C_3H_7Cl	C_3H_7MgCl	77, 153	75	11
$Mg(^3P)$	C_3H_7Cl	$CH_3CH{=}CH_2 +$ $C_3H_7 +$ $n\text{-}C_6H_{14}$	77	35 34	11
Pd	CF_3Br	$(CF_3PdBr)_n$	77		194
	$C_3F_7COCl,$ PEt_3	$C_3F_7COPdCl$		15	
	$(CF_3CO)_2O +$ PEt_3	$cis\text{-}(PEt_3)_2Pd(OCOCF_3)_2$		15	
	$CF_3CF{=}CFCF_3$ $+PEt_3$	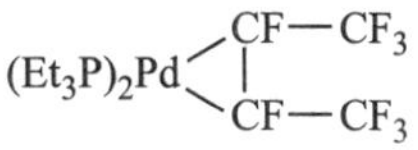			
	$CF_3CF{=}CFCF_3$	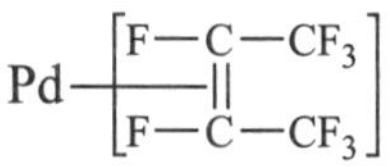			

TABLE 1.9 (Contd.)

Metal	Ligand	Product	T, K	Yield, %	Ref.
Pd	RCOCl, PEt$_3$ R = n-C$_3$F$_7$, CF$_3$, CH$_3$, n-C$_3$H$_7$	(PEt$_3$)$_2$Pd(RCOCl)	77	4–19	197
Mg, Al, Ga, ln	CH$_3$Br	CH$_3$MBr	77		213
Pd	RX, PEt$_3$ R = CF$_3$, C$_2$F$_5$, n-C$_3$F$_7$,CCl$_3$ X = Cl, Br, I	(PEt$_3$)$_2$PdRX + (PEt$_3$)$_2$PdX$_2$	77	0–40	197
Pt	PhCH$_2$Cl + P(OEt)$_3$	(PhCH$_2$){P(OEt)$_3$}$_2$PtCl	77	45	198
	CF$_3$COCl	(CF$_3$CO)Pt(CO)Cl		27	
	CF$_3$COCl, PEt$_3$	(CF$_3$CO)(PEt$_3$)$_2$PtCl		57	
	CF$_3$I, PEt$_3$	(PEt$_3$)$_2$PtI$_2$ + (PEt$_3$)$_2$(CF$_3$)PtI		12 26	
Ni	Simple alkyl- fluoro- and chloroolefins	Ni(olefin)$_2$, n = 1–3	10–77		116, 214
Ni	CH$_2$=CHBr	π-allylnickel halides	77	75	215
Ni, Pd	RX (X = Cl, Br, I; R = C$_6$F$_5$, C$_6$H$_5$, CF$_3$, C$_3$F$_7$, C$_3$H$_7$)	RMX	77		16
Pd	(CF$_3$CO)$_2$O (CF$_3$CO)$_2$O + PPh$_3$	{(CF$_3$CO)$_2$OPd} cis-{(CF$_3$CO)$_2$O·Pd(PPh$_3$)$_2$}	77		16
Mg	n-C$_3$H$_7$I (CH$_3$)$_2$CHBr (CH$_3$)$_3$CBr PhCl CH$_2$=CHBr	C$_3$H$_7$MgI (unsolvated) (CH$_3$)$_2$CHMgBr (CH$_3$)$_3$CMgBr PhMgCl CH$_2$=CHMgBr	77	76 55 5 58 78	74

in a vacuum vessel. During the Mg evaporation ca $20\,cm^3$ of THF (or hexane) was co-deposited (the vessel was immersed in liquid nitrogen). Upon completion of the deposition the vessel was isolated from the vacuum pump and allowed to warm to room temperature. The Mg was isolated by surrounding the vessel with a polyethylene glove bag, flushing with argon, filling the vessel with argon and transferring the Mg–THF to a weighed tube with a stopcock. The Mg–THF slurry could be used directly, or the THF could be pumped off leaving dry active Mg (see [243], p.243).

Zn, Cd + alkyl iodides

Cadmium slurries in diglyme, dioxane, THF, hexane and toluene were prepared by cocondensing about 9 g of Cd vapor with $\approx 60\,cm^3$ of solvent. Colored matrices were formed that turned black upon meltdown. Addition of alkyl iodides to the slurries was followed by warming and then refluxing at reduced pressure [74,206].

Pd + trifluoromethyl bromide

Palladium atoms were co-condensed with trifluoromethyl bromide at 77 K (reaction 1.34), the mixture was warmed slowly to 195 K, then excess substrate was pumped off. Triethylphosphine and CH_2Cl_2 were condensed into the reactor, followed again by warming, stirring, pumping off, washing the solid residue with acetone, filtering, decolorizing, pumping off, and crystallizing the resultant solid from methanol yielding 0.04 g (7%) of white *trans*-bis(triethylphosphine)trifluoromethylpalladium bromide, m.p. 97–98°C [194]:

$$Pd_{at.} + CF_3Br \rightarrow (CF_3PdBr)_n \xrightarrow[-78°C \text{ or } 25°C]{PEt_3} trans\text{-}(Et_3P)_2Pd(CF_3)Br$$

$$(1.34)$$

Al + ethers

The interaction between aluminum atoms in their ground state and a series of ethers was carried out in an inert hydrocarbon matrix (adamantane) at 77 K in the rotating cryostat. Al atoms were deposited at a rate of ca $0.06\,g\,h^{-1}$ by resistively heating the furnace to ca 800°C. Dimethyl ether, diethyl ether, ethylene oxide and 1,3,5-trioxane were used as ligands. The following products have been obtained, among others [245]: CH_3AlOCH_3, $C_2H_5AlOC_2H_5$ and CH_3OCH_2AlH.

Fe + aminodifluorophosphines

About 0.5 g (8.93 mg-atoms) of iron metal was evaporated over a period of 30 min at 0.0001 mm pressure from an alumina-coated tungsten-wound crucible heated electrically. The resulting vapors were co-condensed at 77 K with an excess ($\sim 8\,cm^3$) of dimethylaminodifluorophosphine. Excess ligand was then removed by pumping at 195 K. The resulting black slurry was extracted with hexane and the hexane solution was chromatographed on a Florisil column. Elution of the yellow band with hexane gave 0.8 g (14% yield) of yellow $[(CH_3)_2NPF_2]_5Fe$ [92].

1.4.6 Interaction of metals with polymers

A possibility of using cryosynthesis for complexes of the type **XXXI** is of significance for the preparation of metal-containing polymers [216–220]. Thus, metal-containing polymers of titanium **LII–LIII** [216,218,220], vanadium [216,218,220], chromium [216,218–221], molybdenum [216–218,220], and tungsten [216,218,220] have been synthesized from poly(methylphenylsiloxane). The reactions were carried out by co-condensation of metal vapors and polymers at 240–270 K according to the general scheme (1.35).

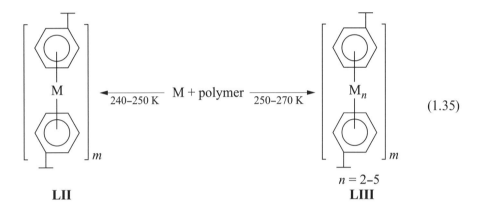

$$\text{(1.35)}$$

$$n = 2\text{–}5$$

LII   **LIII**

The reaction (1.35) takes place in steps. The mono-, di-, and trinuclear polymers [216,218] were isolated for chromium; in the case of molybdenum $n = 1$–5 [217]. Both homo- and heterometallic polymers have been described; they could be obtained by evaporation of different metals, either simultaneously or successively. It should be noted that polystyrene could also be used in this type of transformation [220].

In relation with the reaction (1.35), it is necessary to emphasize the following aspects: first, this transformation opens the way to synthesize coordination metal polymers [222–226]; second, it is possible to obtain metal–graphites [226].

Some reactions of metal atoms with polymers are presented in Table 1.10.

TABLE 1.10
Cryosynthesis using polymers

Metal	Ligand	Product	T, K	Ref.
Ti, V, Cr, Mo, W	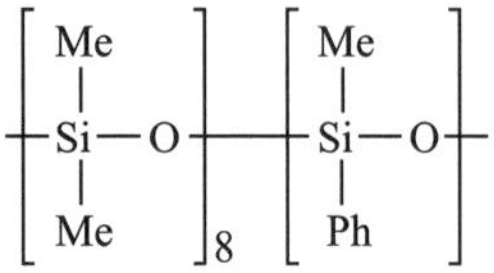 Poly(methyl-phenylsiloxanes)	$Bis(\eta^6$-arene) complexes (arene = SiO(Ph) groups) 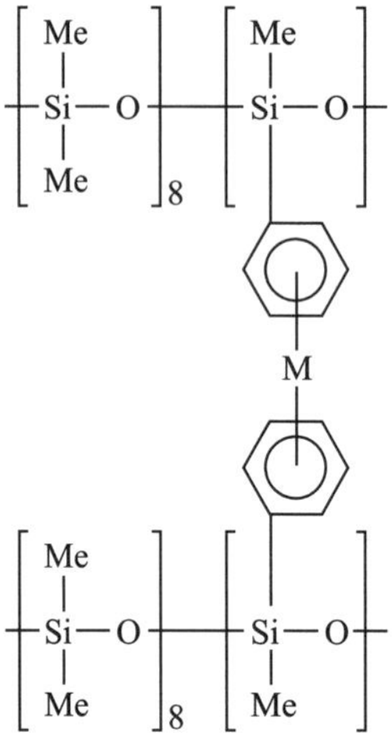	77, 273	6, 227–229

$$\left[\begin{array}{c} \text{Me} \\ | \\ -\text{Si}-\text{O}- \\ | \\ \text{Me} \end{array}\right]_8 \left[\begin{array}{c} \text{Me} \\ | \\ -\text{Si}-\text{O}- \\ | \\ \text{Ph} \end{array}\right]$$

$$\left[\begin{array}{c} \text{Me} \\ | \\ -\text{Si}-\text{O}- \\ | \\ \text{Me} \end{array}\right]_8 \left[\begin{array}{c} \text{Me} \\ | \\ -\text{Si}-\text{O}- \\ | \\ \text{(arene)}-\text{M}-\text{(arene)} \end{array}\right]$$

$$\left[\begin{array}{c} \text{Me} \\ | \\ -\text{Si}-\text{O}- \\ | \\ \text{Me} \end{array}\right]_8 \left[\begin{array}{c} \\ | \\ -\text{Si}-\text{O}- \\ | \\ \text{Me} \end{array}\right]$$

			253	216

Ti + Cr simultaneously

(structure: arene–Ti–arene ... arene–Cr$_{1,2}$–arene polymer)

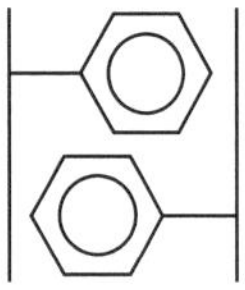

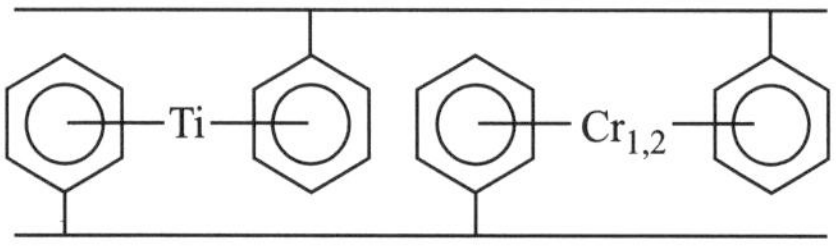

	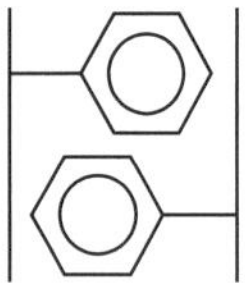	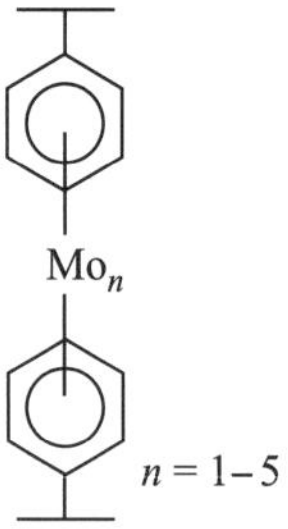	250	217, 219, 230

Mo, Ti, Cr, V

(structure: arene–Mo$_n$–arene, $n = 1–5$)

Experimental procedures

Mo + poly(methylphenylsiloxane) (DC510)

Molybdenum (1 g, 10 mmol) was evaporated from a wire filament and condensed into 100 cm^3 of DC510 contained in a rotary solution reactor at 250 K over a period of 2 h. The formation of products (Table 1.10) was then followed by means of UV/visible spectroscopy [217].

Different metals + poly(methylphenylsiloxane)

Chromium and titanium could be deposited onto the polymer by means of the following reactions: (1) sequential Ti and Cr vapor deposition; (2) simultaneous Ti/Cr vapor deposition; (3) saturation of the phenyl groups on DC510 with Ti vapor followed by reaction with Cr atoms; (4) saturation with Cr vapor followed by reaction with Ti atoms [216]. One of the products of sequential deposition is presented in Table 1.10.

1.5 VAPOR SYNTHESIS OF METAL CHELATES

Practically all metal chelates which have been obtained either by cryosynthesis or by reactions between metal particles and an organic substrate in the gas phase are β-diketonates of transition and nontransition metals [7,41,60]. This is explained by the high stability of β-diketonates and their precursors (β-diketones) in the conditions of gas-phase synthesis at high and low temperatures.

The cryosyntheses of metal acetylacetonates with the general formula M(acac)$_2$, where M = Mn, Cr, Fe, Ni, Pd, Cu, Zn, Sn, Pb, and M(acac)$_3$, where M = Al, Cr, Fe, Dy, Ho, Er, were described in early investigations on co-condensation of atomic metals and acetylacetone, and generalized in reviews [16,41]. The oxidation number of the metal and the amount of molecular hydrogen evolved depend on the synthesis conditions (in particular, on the presence of molecular oxygen and the rate of metal evaporation), the yields being 10–32%.

However, later [60] it proved possible to obtain a series of β-diketonates at 363–543 K (M = Co, Ni, Cu, Cr, Fe, Ti, Zr, Hf) with yields close to quantitative (90–100%), using small metal particles instead of "metal atoms". In spite of a very exotic synthetic route, this method could be used, in the opinion of Mazurenko [60], for industrial production of β-diketonates. It was also noted [60] that the productivity of this process increases sharply when the ligands are used in the form of aerosols. It allows one to introduce significant amounts of β-diketones into the reaction zone with simultaneous regulation of their concentration.

Among the other studies on this aspect, the cryosynthesis of coordination compounds containing radionuclides should be mentioned, in particular, ^{63}Ni acetylacetonate [231].

The feasibility of using cryosynthesis to obtain a wide variety of metal chelates with the coordination kernel MN_2O_2 was established by Kuzharov et al. [232]. Copper and nickel complexes of this type are obtained from the chelating ligands (weak acids) **LIV–LVI** and have the composition ML_2, where LH is the initial ligand system.

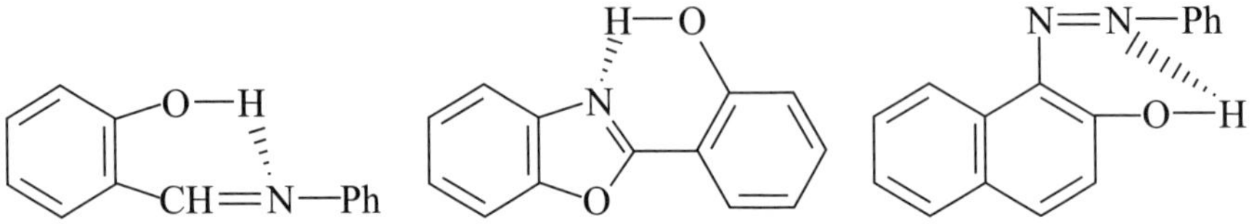

LIV   **LV**   **LVI**

The yields of synthesized products were 10–60%; they were characterized by IR spectroscopy and X-ray diffraction. The complexes obtained are generally identical to the chelates prepared from the same ligands and metal salts. The main advantages of cryosynthesis of metal chelates, in comparison with conventional syntheses, are the following: the time of the process is reduced to a few minutes, no solvents are used, and the products have the precise composition ML_n, whereas, starting from metal salts, the chelates ML_mX_n $(X = Hal^-, NO_3^-, CH_3COO^-$ etc.) could also be obtained.

Experimental procedures

Atoms of a series of transition and f-metals (Cu, Zn, Ni, Cr, Fe, Sn, Pd, Er, Dy, Ho) were co-condensed with acetylacetone previously frozen in the reactor walls at 77 K. The yields were 10–32%. The proposed mechanism of β-diketonate formation is represented by scheme (1.36) [41]:

$$M + RY \longrightarrow M\begin{matrix} R \\ \diagup \\ \diagdown \\ Y \end{matrix} \xrightarrow{RY} \begin{matrix} Y \\ \diagdown \\ \diagup \\ Y \end{matrix} M \begin{matrix} R \\ \diagup \\ \diagdown \\ R \end{matrix} \longrightarrow M\begin{matrix} R \\ \diagup \\ \diagdown \\ R \end{matrix} + Y_2 \qquad (1.36)$$

Using small metal particles of Cr, Fe, Co, Ni, Ti, Zr, Hf (instead of metal atoms) carried by an inert gas into a reactor (Fig. 1.7), it is possible to increase yields of metal β-diketonates to 90–100% [60]. If the metal particles are in the 0.1–0.3 mm size range and their specific surface is between 50 and 100 $m^2 m^{-1}$, the time for the complete reaction is reduced to a fraction of a second. The product output increases sharply (up to 10–60 $g\,min^{-1}$ of β-diketonate) when the β-diketone is introduced to the reactor as an aerosol. The data on some synthesized metal β-diketonates are presented in Table 1.11.

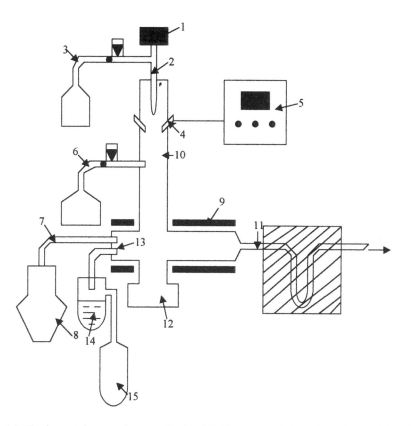

Fig. 1.7. Equipment for gas-phase synthesis of β-diketonates: 1, Metal powder container; 2, metal powder sprayer; 3, inert gas feeding system; 4, plasmotrone; 5, apparatus for the control of plasmotrones; 6, gas feed system for powder hardening; 7, liquid-nitrogen vapor feed sprayer; 8, liquid-nitrogen container; 9, thermostat; 10, reactor; 11, refrigerator to catch metal β-diketonate; 12, unreacted metal accumulator; 13, β-diketone aerosol formation sprayer; 14, saturator; 15, gas feeding system for β-diketone aerosol formation [60].

TABLE 1.11

Synthesis conditions of metal β-diketonates in the gas phase [60]

Metal, β-diketone	Synthesis conditions					
	Dimensions of metal powder particles, μm	Gas-bearer velocity, $dm^3 h^{-1}$	Metal feed rate, g min^{-1}	β-Diketone feed rate, g min^{-1}	Temp., °C	Yield, %
Cr(III) Hacac	0.25–0.30	0.12–0.25	3.0	16.4	270	98.6
Cr(III) HTFA	0.25–0.30	0.12–0.25	1.0	8.9	180	100
Cr(III) HHFA	0.25–0.30	0.12–0.25	1.0	12.0	90	100
Fe(III) Hacac	0.20–0.30	0.20–0.30	3.5	18.6	195	100
Fe(III) HTFA	0.20–0.30	0.20–0.30	1.0	8.2	150	100
Fe(III) HHFA	0.20–0.30	0.20–0.30	0.5	5.5	120	100
Co(II) Hacac	0.20–0.30	0.12–0.25	2.8	9.4	160	100
Co(II) HTFA	0.20–0.30	0.12–0.25	2.0	6.8	140	100
Co(II) HHFA	0.20–0.30	0.12–0.25	0.5	1.7	110	100
Ni(II) Hacac	0.20–0.30	0.12–0.25	3.0	10.2	180	100
Ni(II) HTFA	0.20–0.30	0.12–0.25	1.5	5.0	150	100
Ni(II) HHFA	0.20–0.30	0.12–0.25	3.2	10.8	120	100
Cu(II) Hacac	0.20–0.30	0.12–0.25	2.5	8.5	220	100
Cu(II) HTFA	0.20–0.30	0.5	12.0	57.8	140	92
Ti(IV) Hacac	0.25	0.3	5.2	5.0	200	90
Zr(IV) Hacac	0.25	0.5	5.2	24.5	230	100
Hf(IV) Hacac	0.25	0.5	12.0	27.5	190	100

1.6 CONCLUSIONS

As was shown in this chapter, cryosynthesis with the participation of atomic metals at low temperatures, and "pure" gas-phase reactions without a condensed phase at high temperatures, hold many possibilities for the synthesis of various types of complex, especially π-complexes of aromatic and cyclodiene ligands. The main advantages of this method are the following: it is possible (and frequently it is a unique way) to prepare desirable metal π-complexes without solvent molecules or acidic radicals and to reduce the time of synthesis. Noteworthy among the π-complexes are those with η^6-coordinated nitrogen-containing heterocycles [183–185,233,234], which it has been impossible to obtain by conventional synthetic routes. In this respect we note that a series of hetarene compounds ML_n ($L = \alpha, \alpha'$-bipy, C_5H_5P), obtained by the conventional chemical route, could be attributed to π-complexes [58,235]. Thus, as a result of the interaction of bis(1,5-cyclooctadiene)nickel with C_5H_5P, the coordination compound $(C_5H_5P)_4Ni$ **LVII** was isolated, which, however, has the structure of a σ-complex according to X-ray diffraction data [235].

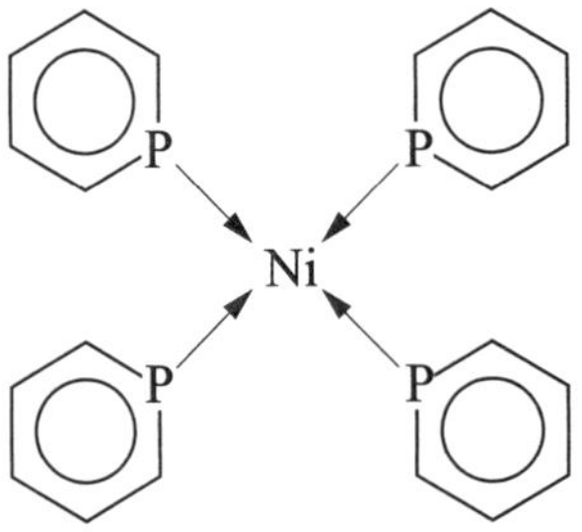

LVII

The η^1-complexes of C_5H_5P with chromium $[(\eta^1\text{-}C_5H_5P)_6Cr]$ and iron $[(\eta^1\text{-}C_5H_5P)_5Fe]$ have also been described [161]. However, the attempts to isolate similar compounds for pyridine were unsuccessful, although their formation in solutions was established [235].

Direct gas-phase synthesis permits one to produce extremely pure chemical substances. For example, the usual synthesis of bis(ethylbenzene)chromium by Fisher's method leads to the formation of many homologues with different numbers of alkyl substituents; use of cryosynthesis produces an individual compound [16]. It is very important in the preparation of metal films or standard substances with constant characteristics: melting and boiling points, vapor pressure, viscosity, density, etc. Another recent important application of metal-vapor chemistry [74] is the formation of intrazeolite metal clusters, where small metal clusters are deposited in the supercages of zeolite. Other detailed applications of the cryosynthesis are presented in the same monograph [74].

However, this method has also the following significant disadvantages: it is necessary to use special high-vacuum equipment, which is not available everywhere; and yields are frequently lower in comparison with the conventional methods.

Taking into consideration all the discussion above and the great interest in developing metal chelates of various organic ligands (for example, [236–239]), we have to regret that gas-phase synthesis on the basis of chelate-forming ligands is very limited and data on the cryosynthesis of adducts of organic ligand systems with N,P,O,S-donor centers (L) [21,236] are practically absent. This last fact is strange, since the formation of adducts of the type $MX_n \cdot mL$ is theoretically possible using metal, ligand and halogen vapors in these conditions. Perhaps the adducts of metal chelates could also be obtained, for example those of acetylacetonates with N-bases, $M(acac)_2 \cdot mL$. However, it is necessary to take into account the possibility of dissociation of either the adducts or Lewis acids (which are used to prepare the adducts) in gas-phase conditions.

In our opinion, in the gas phase it is necessary to carry out the syntheses of compounds that are accessible only with difficulty by conventional chemical methods, and such reactions as complex formation, in the gas phase, where higher yields of metal complexes could be expected in comparison with the conventional methods.

The most recent review articles and monographs on cryosynthesis and related areas (matrix isolation) are [19–21,78,93,240–242] and [25b,27,74, 243], respectively. A wide range of possibilities have been opening for the study of reaction products using ion-cyclotron resonance ([28,29,244] and references therein).

ACKNOWLEDGMENT

The authors are very grateful to Professor Kenneth J. Klabunde (Kansas State University) for the critical revision of this chapter and the permission to reproduce illustrations from his books.

REFERENCES

1. Skell, P.S.; Wescott, L.D. *J. Am. Chem. Soc.* **85**(7), 1023 (1963).
2. Skell, P.S.; Wescott, L.D.; Goldstein, I.P.; Engel, R.R. *J. Am. Chem. Soc.* **87**(13), 2829 (1965).
3. Timms, P.L. *J. Chem. Soc., Chem. Commun.* **18**, 1033 (1969).
4. a) Kargin, V.A.; Kabanov, V.A.; Zubov, V.P. *Visokomol. Soedin.* **2**(2), 303 (1960); b) Kargin, V.A.; Kabanov, V.A.; Zubov, V.P. *Visokomol. Soedin.* **3**(3), 426, (1961).
5. Timms, P.L. *Adv. Inorg. Chem. Radiochem.* **14**, 121 (1972).
6. Timms, P.L. *Acc. Chem. Res.* **6**(4), 118 (1973).
7. Timms, P.L. *Angew. Chem.* **87***(5), 295 (1975); Agew. Chem. Int. Ed. Engl.* **14**(5), 273, (1975).

8. Timms, P.L.; Turney, T.W. *Adv. Organomet. Chem.* **15**(1), 53 (1977).

9. Skell, P.S.; Havel, J.I.; McGlinchey, M.J. *Acc. Chem. Res.* **6**(3), 97 (1973).

10. Timms, P.L. *Transition Met. Chem., Proc. 1980 Workshop.* 23 (1981).

11. Skell, P.S.; McGlinchey, M.J. *Angew. Chem.* **8**(5), 215 (1975); *Int. Ed. Engl.* **14**(5), 195 (1975).

12. Klabunde, K.J. *Angew. Chem.* **87**(5), 309 (1975); *Int. Ed. Engl.* **14**(5), 287 (1975).

13. Klabunde, K.J. *Acc. Chem. Res.* **8**(12), 393 (1975).

14. Zoellner, R.W.; Klabunde, K.J. *Chem. Rev.* **84**(6), 545 (1984).

15. Power, W.J.; Ozin, G.A. *Adv. Inorg. Chem. Radiochem.* **23**, 80 (1980).

16. Domrachev, G.A.; Zinoviev, V.D. *Usp. Khim.* **47**(4), 679 (1978).

17. Domrachev, G.A.; Zakharova, L.N.; Shevelev, Yu.A. *Usp. Khim.* **54**(8), 1260 (1985).

18. Sergeev, G.B. *Zh. Vses. Khim. Obshch. D.I.Mendeleeva.* **35**(5), 566 (1990).

19. Garnovskii, A.D.; Kharisov, B.I.; Gójon-Zorrilla, G.; Garnovskii, D.A. *Russ. Chem. Rev. (Engl. Trans.)* **64**(3), 201 (1995).

20. Gójon-Zorrilla, G.; Kharisov, B.I.; Garnovskii, A.D. *Rev. Soc. Quím. Méx.* **40**(3), 131 (1996).

21. Garnovskii, A.D.; Kharisov, B.I.; Gójon-Zorrilla, G.; Garnovskii, D.A.; Burlov, A.S. *Koord. Khim.* **23**(4), 243 (1997).

22. Blackborow, J.R.; Young, D. *Metal Vapor Synthesis in Organometallic Chemistry.* Berlin: Springer-Verlag Chemie (1979).

23. *Cryochemistry* (Eds: Moskovits, M.; Oxin, G.A.). New York: Wiley (1976).

24. *Inorganic Synthesis*, Vol. 19, Chapt. 2. New York: McGraw Hill (1979).

25. a) Klabunde, K.J. *Chemistry of Free Atoms and Particles.* New York: Academic Press (1980); b) Klabunde, K.J. *Free Atoms, Clusters, and Nanoscale Particles.* San Diego, CA: Academic Press (1994).

26. *Gas Phase Inorganic Chemistry* (Ed: Russel, D.H.). New York: Plenum Press (1989).

27. *Direct Synthesis of Coordination Compounds* (Ed: Skopenko, V.V.). Kiev: Ventury (1997), 186 pp.

28. Eller, K.; Schwarz, H. *Chem. Rev.* **91**, 1121 (1991).

29. Eller, K. *Coord. Chem. Rev.* **126**(1–2), 93 (1993).

30. Huber, H.; Kündig, E.P.; Moskovits, M.; Ozin, G.A. *J. Am. Chem. Soc.* **95**(2), 332 (1973).

31. Pochekutova, T.S.; Khamilov, V.K.; Domrachev, G.A.; Zhuk, B.V. *Zh. Org. Khim.* **23**(10), 2156 (1987).

32. Buckner, S.W.; Freiser, B.S. *Polyhedron* **7**(16/17), 1583 (1988).

33. Blitz, M.A.; Mitchell, S.A.; Hackett, P.A. *J. Phys. Chem.* **95**(22), 8719 (1991).

34. Carrol, J.J.; Weisshaar, J.C. *J. Am. Chem. Soc.* **115**(2), 800 (1993).

35. Carrol, J.J.; Hang, K.L.; Weisshaar, J.C. *J. Am. Chem. Soc.* **115**(15), 6962, (1993).

36. Bitter, D.; Weisshaar, J.C. *J. Am. Chem. Soc.* **112**(17), 6425 (1990).

37. Pimentel, G.C. *Angew. Chem.* **87**(5), 220 (1975); *Angew. Chem. Int. Ed. Engl.* **14**(4), 199 (1975).

38. Purcell, K.E.; Kotz, J.C. *Inorganic Chemistry.* Philadelphia: Saunders Golden Sunburst Series (1977), pp. 813–825, 861, 862, 879, 880.

39. Domrachev, G.A.; Andreev, G.I.; Shevelev, Yu.A.; Nemtsov, B.E. *Zh. Prikl. Khim.* **57**(6), 1266 (1984).

40. Ivanov, A.A.; Mikhailov, G.M.; Kukushkin, V.I. USSR Patent 1607929 (1990).
41. Von Gustoff, E.A.K.; Jenicke, O.; Wolfbeis, O.; Eady, C.R. *Angew. Chem.* **87**(5), 300 (1975); *Angew Chem. Int. Ed. Engl.* **14**(5), 278 (1975).
42. Graves, V.; Lagowski, J.J. *Inorg. Chem.* **15**(3), 577 (1976).
43. a) Benfield, F.W.S.; Green, M.L.H.; Oigen, J.S.; Young, D. *J. Chem. Soc., Chem. Commun.* 866 (1973); b) Elschenbroich, C.; Hoppe, S.; Mertz, B. *Chem. Ber.* **126**(2), 399 (1993).
44. Domenico, M.; Timms, P.L. *J. Organomet. Chem.* **253**(1), 12 (1983).
45. Wittchell, S.A.; Hackett, P.A. *J. Phys. Chem.* **93**(11), 7822 (1990).
46. Ball, D.W.; Pong, R.G.; Kafari, H. *J. Am. Chem. Soc.* **115**(7), 2864 (1993).
47. Blackborow, J.R.; Feldhoff, U.; Grevels, F.-W. *J. Organomet. Chem.* **173**(2), 253 (1979).
48. Ozin, G.A.; Haddleton, D.M.; Gil, C.J. *J. Phys. Chem.* **92**(9), 6710 (1989).
49. Skell, P.S.; Havel, J.J.; Williams-Smith, D.L.; McGlinchey, M.J. *J. Chem. Soc., Chem. Commun.* **19**, 1098 (1972).
50. Kündig, E.P.; Timms, P.L.; Kelly, B.A.; Woodward, P. *J. Chem. Soc., Dalton Trans.* **9**, 901 (1983).
51. Cloke, F.G.N.; Green, M.L. *J. Chem. Soc., Dalton Trans.* **18**, 1938 (1981).
52. Cloke, F.G.N.; Lappert, N.F.; Lawless, G.A.; Swain, A.C. *J. Chem. Soc., Chem. Commun.* **21**, 1667 (1987).
53. Brean, J.G.; Cloke, F.G.N.; Sameh, A.A.; Salkin, A. *J. Chem. Soc., Chem. Commun.* **21**, 1668 (1987).
54. Cloke, F.G.N.; Courtney, K.A.E.; Sameh, A.A.; Swain, A.C. *Polyhedron* **8**(13/14), 1641 (1989).
55. Brunner, H.; Koch, H. *Chem. Ber.* **115**(1), 65 (1982).
56. Yurieva, L.P.; Zaitseva, N.N.; Uralets, I.A.; Peregudova, S.I.; Kravtsov, D.N.; Vasilkov, A.Yu.; Sergeev, V.A. *Metalloorg. Khim.* **3**(40), 783 (1990).
57. Simons, A.H.; Riley, P.E.; Davis, R.F.; Lagowski, D. *J. Am. Chem. Soc.* **98**(4), 1044 (1976).
58. Groshens, T.G.; Henne, B.; Bartak, D.; Klabunde, K.J. *Inorg. Chem.* **20**, 3629 (1981).
59. Gardenas-Treviño, G.; Klabunde, K.J.; Dale, E.B. *Langmuir* **3**, 986 (1987).
60. Mazurenko, E.A. Dr. Hab. Thesis. Kiev: IONH Ukr.SSR (1987).
61. Starowiewski, K.B.; Klabunde, K.J. *Appl. Organomet. Chem.* **3**(2), 219 (1989).
62. Razuvaev, G.A.; Domrachev, G.A.; Zinoviev, V.D. *Dokl. Akad. Nauk SSSR.* **223**(3), 617 (1975).
63. Timms, P.L. *J. Chem. Soc. A.* **13**, 2526 (1970).
64. Kündig, E.P.; Moskovits, M.; Ozin, G.A. *Can. J. Chem.* **51**, 2710 (1973).
65. Busby, R.; Klotzbücher, W., Ozin, G.A. *Inorg. Chem.* **16**(4), 822 (1977).
66. Bochmann, M. *Organometallics 1: Complexes with Transition Metal–Carbon σ-Bonds.* Oxford: Oxford University Press (1993), pp. 9–14, 39.
67. Huber, H.; Klotzbücher, W.; Ozin, G.A.; Vander Voet, A. *Can. J. Chem.* **51**, 2722 (1973).
68. Ozin, G.A.; Klotzbücher, W. *J. Am. Chem. Soc.* **97**(14), 3965 (1975).
69. Garnovskii, A.D. *Koord. Khim.* **14**(5), 579 (1988).
70. McIntosh, D.F.; Ozin, G.A.; Messmer, R.P. *Inorg. Chem.* **20**, 3640 (1981).
71. Chenier, J.H.B.; Hampson, C.A.; Howard, J.A.; Mile, B. *J. Phys. Chem.* **92**, 2745 (1988).

72. Chenier, J.H.B.; Hampson, C.A.; Howard, J.A.; Mile, B. *J. Phys. Chem.* **93**, 114 (1989).

73. Chenier, J.H.B.; Histed, M.; Howard, J.A.; Joly, H.A.; Morris, H.; Mile, B. *Inorg. Chem.* **28**, 4114 (1989).

74. Cintas, P. *Activated Metals in Organic Synthesis.* Boca Raton: CRC Press (1993), pp. 12–44.

75. Cesaro, S.N.; Dobos, S. *Microchim. Acta, Suppl.* **14** (Progress in Fourier Transform Spectroscopy), 387 (1997).

76. Kündig, E.P.; Moskovits, M.; Ozin, G.A. *Can. J. Chem.* **51**, 2737 (1973).

77. Wagner, F.S. *Kirk-Othmer Encycl. Chem. Technol.*, 3rd edn. **4**, 794 (1978).

78. Timney, J.A. *J. Organomet. Chem.* **25**, 150 (1996).

79. Yarborough, L.W.; Calder, G.V.; Verkade, J.G. *J. Chem. Soc., Chem. Commun.* **5**, 705 (1972).

80. Huber, H.; Ozin, G.A.; Power, W.J. *Inorg. Chem.* **16**(9), 2234 (1977).

81. Kushto, G.P.; Souter, P.F.; Andrews, L.; Neurock, M.J. *J. Chem. Phys.* **106**(14), 5894 (1997).

82. Klotzbücher, W.E.; Petrukhina, M.A.; Sergeev, G.B. *J. Phys. Chem., A* **101**, 4548 (1997).

83. García, P.J.; Novaro, O. *Rev. Méx. Fís.* **43**(1), 130 (1997).

84. Galan, F.; Fouassier, M.; Tranquille, M.; Mascetti, J.; Papai, I. *J. Phys. Chem. A* **101**(14), 2626 (1997).

85. Marquardt, R.; Sander, W.; Laue, T.; Hopf, H. *Liebigs Ann. Chem.* **12**, 2039 (1996).

86. Widenhoefer, R.A. *Diss. Abstr. Int. B.* **55**(11), 4846 (1995).

87. a) Fedrigo, S.; Haslett, T.L.; Moskovits, M.J. *J. Am. Chem. Soc.* **118**(21), 5083 (1996); b) Fedrigo, S.; Haslett, T.L.; Moskovits, M.J. *Z. Phys. D.: At. Mol. Clusters* **40**(1–4), 99 (1997).

88. Chertihin, G.V.; Andrews, L.; Bauschlicher, C.W.Jr. *J. Phys. Chem. A* **101**(22), 4026 (1997).

89. Fontijn, A.; Blue, A.S.; Narayan, A.S.; Bajaj, P.N. *Combust. Sci. Technol.* **101**(1–6), 59 (1994).

90. Ball, D.W.; Hauge, K.H.; Margrave, J.L. *High Temp. Sci.* **25**(1), 95 (1988).

91. Park, M.; Hauge, K.H.; Margrave, J.L. *High Temp. Sci.* **25**(1), 1 (1988).

92. Chang, M.; King, R.B.; Newton, M.G. *J. Am. Chem. Soc.* **100**(3), 998 (1978).

93. Schneider, J.J. *Angew. Chem., Int. Ed. Engl.* **35**(10), 1068 (1996).

94. McIntosh, D.; Ozin, G.A. *Inorg. Chem.* **16**(1), 59 (1977).

95. Hanlan, A.J.L.; Ozin, G.A. *Inorg. Chem.* **16**(11), 2848 (1977).

96. Hanlan, A.J.L.; Ozin, G.A. *Inorg. Chem.* **16**(11), 2857 (1977).

97. Midletton, R.; Hall, J.R.; Simpson, S.R.; Tominson, C.H.; Timms, P.L. *J. Chem. Soc., Dalton Trans.* **1**, 120 (1973).

98. Huber, H.; Ford, T.A.; Klotzbücher, W.E.; Ozin, G.A. *J. Am. Chem. Soc.* **98**(11), 3176 (1976).

99. Ozin, G.A.; Moskovits, M.; Kündig, P.; Huber, H. *Can. J. Chem.* **50**, 2385 (1972).

100. Rack, J.J. *Univ. Microfilms Int.* Colorado State Univ., Fort Collins, CO, 115 pp. (Order No. DA9638706.1996).

101. Chenier, J.H.B.; Howard, J.A.; Joly, H.A.; Mile, B.; Tomiento, M. *Can. J. Chem.* **67**, 655 (1989).

102. Ozin, G.A.; Hanlan, A.J.L. *Inorg. Chem.* **18**, 8 (1979).
103. Ford, T.A.; Huber, H.; Klotzbücher, W.E.; Moskovits, M.; Ozin, G.A. *Inorg. Chem.* **15**(7), 1666 (1976).
104. Godber, J.; Huber, H.; Ozin, G.A. *Inorg. Chem.* **25**, 2909 (1986).
105. McIntosh, D.; Ozin, G.A. *J. Am. Chem. Soc.* **98**(11), 3167 (1976).
106. Timms, P.L. *La Reserche* **105**, 1090 (1979).
107. Huber, H.; Ozin, G.A. *Inorg. Chem.* **16**(1), 64 (1977).
108. Huber, H.; McIntosh, D.; Ozin, G.A. *Inorg. Chem.* **16**(5), 975 (1977).
109. Ozin, G.A.; Hanlan, A.J.L. *Inorg. Chem.* **18**(9), 2390 (1979).
110. Chenier, J.H.B.; Howard, J.A.; Joly, H.A.; Mile, B.; Timms, P.L. *J. Chem. Soc., Chem. Commun.* 581 (1990).
111. Ozin, G.A.; McIntosh, D.; Mitchell, S.A. *J. Am. Chem. Soc.* **103**, 1574 (1981).
112. Chung, C.; Lagow, R.J. *J. Chem. Soc., Dalton Trans.* 1078 (1972).
113. Ozin, G.A.; Huber, H.; McIntosh, D. *Inorg. Chem.* **17**(6), 1472 (1978).
114. Hooker, P.D.; Timms, P.L. *J. Chem. Soc., Chem. Commun.* 158 (1988).
115. Hanlan, A.J.L.; Ozin, G.A.; Power, W.I. *Inorg. Chem.* **17**(20), 3648 (1978).
116. Ozin, G.A.; Power, W.I. *Inorg. Chem.* **17**(10), 2836 (1978).
117. Skell, P.S.; Havel, J.J. *J. Am. Chem. Soc.* **93**(24), 6687 (1971).
118. Howard, J.A.; Joly, H.A.; Mile, B. *J. Phys. Chem.* **96**(2), 1233 (1992).
119. Alikhani M.E.; Bouteiller, Y.; Silvi, B. *J. Phys. Chem.* **100**(40), 16092 (1996).
120. Roszak, S.; Balasubramanian, K. *Chem. Phys. Lett.* **231**(1), 18 (1994).
121. Huber, H.; Ozin, G.A.; Power, W.I. *J. Am. Chem. Soc.* **98**(21), 6508 (1976).
122. Huber, H.; Ozin, G.A.; Power, W.I. *Inorg. Chem.* **16**(5), 979 (1977).
123. Ozin, G.A.; Power, W.I. *Inorg. Chem.* **16**(1), 212 (1977).
124. McIntosh, D.; Ozin, G.A. *J. Organomet. Chem.* **121**, 127 (1976).
125. Nikolaas, J.R.; Van Eikema, H.; Bickelhaupt, F.; Klumpp, G.W. *Angew. Chem. Int. Ed. Engl.* **27**(8), 1083 (1988).
126. Ozin, G.A.; Huber, H.; McIntosh, D. *Inorg. Chem.* **16**(12), 3070 (1977).
127. Skell, P.S.; McGlinchey, J. *New Synth. Methods* **3**, 7 (1975).
128. Atkins, M.; Mackenzie, R.; Timms, P.L.; Turney, T.W. *J. Chem. Soc., Chem. Commun.* 764 (1975).
129. Chenier, J.H.B.; Joly, H.A.; Howard, J.A.; Mile, B. *J. Chem. Soc., Faraday Trans.* **86**(19), 3329 (1990).
130. Blackborow, J.R.; Eady, C.R.; Von Gustorf, E.A.K.; Scrivanti, A.; Wolfbeis, O. *J. Organomet. Chem.* **111**, C3 (1976).
131. Von Gustorf, E.A.K.; Jaenicke, O.; Polansky, O.E. *Angew. Chem.* **84**(11), 547 (1972); *Angew. Chem. Int. Ed. Engl.* **11**(10), 533 (1972).
132. Williams-Smith, D.L.; Wolf, L.R.; Skell, P.S. *J. Am. Chem. Soc.* **94**, 4042 (1972).
133. Skell, P.S.; Vam Dam, E.M.; Silvon, M.P. *J. Am. Chem. Soc.* **96**, 626 (1974).
134. Chenier, J.H.B.; Howard, J.A.; Tse, J.S.; Mile, B. *J. Am. Chem. Soc.* **107**(25), 7290 (1985).
135. Domrachev, G.A.; Zinoviev, V.D. *Izvest. Akad. Nauk SSSR, Ser. Khim.* **6**, 1429 (1976).
136. Nesmeyanov, A.N.; Razuvaev, G.A.; Materikova, R.B. et al. *Zh. Obshch. Khim.* **48**(9), 2132 (1978).

137. Barreta, A.; Chong, K.S.; Cloke, F.G.N.; Feigenbaum, A.; Green, M.L.H. *J. Chem. Soc., Dalton Trans.* 861 (1983).

138. Schneider, J.J.; Specht, U.; Goddard, R.; Krueger, C. *Chem. Ber./Recl.* **130**(2), 161 (1997).

139. Schneider, J.J.; Krueger, C. *Chem. Ber.* **125**(4), 843 (1992).

140. Tacke, M.; Teuber, R. *J. Mol. Struct.* **408**, 507 (1997).

141. Tacke, M. *Organometallics* **13**(10), 4124 (1994).

142. Zimmerman, G.I.; Hall, L.W.; Sneddon, L.G. *Inorg. Chem.* **19**(12), 3642 (1980).

143. Knoch, F.; Kremer, F.; Schmidt, U.; Zenneck, U.; Le Floch, P.; Mathey, F. *Organometallics* **15**(14), 2713 (1996).

144. Timms, P.L.; Turney, T.W. *J. Chem. Soc., Dalton Trans.* **20**, 2021 (1976).

145. Starks, D.F.; Streitweiser, A. *J. Am. Chem. Soc.* **95**(19), 3423 (1973).

146. Kolesnikov, S.P.; Dobson, J.E.; Skell, P.S. *J. Am. Chem. Soc.* **100**(3), 999 (1978).

147. Skell, P.S.; Williams-Smith, D.L.; McGlinchey, M.J. *J. Am. Chem. Soc.* **95**(10), 3337 (1973).

148. Bogdanovich, B.; Croener, M.; Wilke, G. *Annalen* **699**, 1 (1966).

149. Baretta, A.; Chong, K.S.; Cloke, F.G.N.; Feigenbaum, A.; Green, M.L.N. *J. Chem. Soc., Dalton Trans.* 861 (1983).

150. Hanlan, A.J.L.; Ugolick, R.C.; Fulcher, J.G.; Togashi, S.; Bocarsly, A.B.; Gladysz, J.A. *Inorg. Chem.* **19**(6), 1543 (1980).

151. Klabunde, K.J. *Am. Lab.* 35 (1975).

152. Klabunde, K.J.; Groshens, T.; Brezinski, M.; Kennellg, W. *J. Am. Chem. Soc.* **100**(3), 4437 (1978).

153. Mackenzie, R.; Timms, P.L. *J. Chem. Soc., Chem. Commun.* 650 (1974).

154. Timms, P.L.; King, R.B. *J. Chem. Soc., Chem. Commun.* **20**, 898 (1974).

155. Shevelev, Yu.A.; Dodonov, M.V.; Yurieva, L.I.; Domrachev, G.A.; Zaitseva, N.I. *Izvest. Akad. Nauk SSSR, Ser. Khim.* **6**, 1414 (1981).

156. Shevelev, Yu.A.; Cherepnov, V.L.; Domrachev, G.A.; Smirnov, V.K. et al. *Dokl. Akad. Nauk SSSR.* **289**(3), 640 (1986).

157. Lynch, V.M.; Yoon, M.O.; Lagowsky, J.J.; Davis, B.E. *Acta Crystallogr., Sect. C* **46**, 1094 (1990).

158. Domrachev, G.A.; Shevelev, Yu.A.; Andreev, I.G.; Makarenko, N.P. *Chemistry of Elementoorganic Compounds.* Gorkii: Nauka (1987), p. 15.

159. Kündig, E.P.; Timms, P.L. *J. Chem. Soc., Dalton Trans.* **9**, 991 (1980).

160. Elschenbroich, C.; Kuehlkamp, P.; Koch, J.; Behrendt, A. *Chem. Ber.* **129**(7), 871 (1996).

161. a) Elschenbroich, C.; Nowotny, M.; Kroker, J.; Behrendt, A.; Massa, W.; Wocadlo, S. *J. Organomet. Chem.* **459**(1–2), 157 (1993); b) Arnold, P.L.; Cloke, F.G.N.; Hitchcock, P.B. *J. Chem. Soc., Chem. Commun.* 481 (1997).

162. Elschenbroich, C.; Schneider, J.; Burdolf, H. *J. Organomet. Chem.* **391**(2), 195 (1990).

163. Elschenbroich, C.; Schneider, J.; Massa, W.; Baum, G.; Mellinghoff, H. *J. Organomet. Chem.* **355**(1–3), 163 (1988).

164. Elschenbroich, C.; Koch, J.; Schneider, J.; Spangenberg, B.; Schiess, P. *J. Organomet. Chem.* **317**(1), 41 (1986).

165. Elschenbroich, C.; Schneider, J.; Prinzbach, H.; Fessner, W.D. *Organometallics* **5**(10), 2091 (1986).

166. Elschenbroich, C.; Spangenberg, B.; Mellinghoff, H. *Chem. Ber.* **117**(10), 3165 (1984).

167. Elschenbroich, C.; Moeckel, R.; Bilger, E. *Z. Naturforsch., Tiel B: Anorg. Chem., Org. Chem.* **39B**(3), 375 (1984).

168. Elschenbroich, C.; Koch, J. *J. Organomet. Chem.* **229**(2), 139 (1982).

169. Zenneck, U. *Angew. Chem., Int. Ed. Engl.* **34**(1), 53 (1995).

170. Green, M.L.; Joyner, D.S.; Wallis, J.M. *J. Chem. Soc., Dalton Trans.* **11**, 2823 (1987).

171. Domrachev, G.A.; Zakharov, L.N.; Shevelev, Yu.A.; Razuvaev, G.A. *Zh. Strukt. Khim.* **27**(2), 14 (1986).

172. Domrachev, G.A. *Problems of the Stability of Organometallic Compounds in the Processes of Their Synthesis and Decomposition.* Moscow: Nauka (1985), p. 138.

173. Mattar, S.M.; Sammynaiken, R. *J. Chem. Phys.* **106**, 1080 (1997).

174. Mattar, S.M.; Sammynaiken, R. *J. Chem. Phys.* **106**, 1094 (1997).

175. Kuznetsov, N.T.; Sevastianov, V.G.; Filatov, I.Yu.; Zakharov, L.N.; Domrachev, G.A. *Visokochist. Veshestva* 42 (1989).

176. Ketkov, S.Yu.; Domrachev, G.A. *Optika i Spektroskopia* **67**(2), 475 (1989).

177. Ketkov, S.Yu.; Domrachev, G.A. *Chugaev Conference on Coordination Compounds.* Minsk (1990), Part 3, p. 484.

178. Ketkov, S.Yu.; Domrachev, G.A. *Metalloorg. Khim.* **3**(4), 957 (1990).

179. Domrachev, G.A.; Ketkov, S.Yu.; Shevelev, Yu.A. et al. *Metalloorg. Khim.* **3**(4), 827 (1990).

180. Tacke, M. *Chem. Ber.* **129**(11), 1369 (1996).

181. Tacke, M. *Chem. Ber.* **128**(1), 91 (1995).

182. Yao, Z.; Klabunde, K.J.; Asirvatham, A.S. *Inorg. Chem.* **34**, 5289 (1995).

183. Riley, P.E.; Davis, R.E. *Inorg. Chem.* **15**, 2735 (1976).

184. Wucherer, E.J.; Muetterties, E.L. *Organometallics* **6**(8), 1691 (1987).

185. Elschenbroich, C.; Koch, J.; Kroker, J.; Wunsch, M.; Massa, W.; Baum, G.; Stork, G. *Chem. Ber.* **121**(9), 1983 (1988).

186. Fisher, E.O.; Ofele, K. *Z. Naturforsch., Teil B* **14**, 736 (1959).

187. Fisher, E.O.; Ofele, K. *J. Organomet. Chem.* **8**(1), 5 (1967).

188. Bogdanov, G.M.; Bundel, Yu.G. *Khim. Geterotsikl. Soedin.* **9**, 1155 (1983).

189. Sadimenko, A.P.; Garnovskii, A.D.; Osipov, O.A.; Sheinker, V.N. *Khim. Geterotsikl. Soedin.* **10**, 1299 (1983).

190. a) Sadimenko, A.P.; Garnovskii, A.D.; Retta, N. *Coord. Chem. Rev.* **126**(1–2), 237 (1993); b) *Advances in Heterocyclic Chemistry* (Ed.: Katritzky, A.R.), Vol. 72. New York: Academic Press (1999).

191. Klabunde, K.J. *New Synth. Methods* **3**, 135 (1975).

192. Klabunde, K.J.; Efner, H.F. *Inorg. Chem.* **14**(4), 789 (1975).

193. Anderson, B.B.; Behrens, C.L.; Radonovich, L.G.; Klabunde, K.J. *J. Am. Chem. Soc.* **98**(17), 5390 (1976).

194. Klabunde, K.J.; Low, J.Y.F.; Efner, H.F. *J. Am. Chem. Soc.* **96**(6), 1984 (1974).

195. Klabunde, K.J.; Efner, H.F. *J. Fluorine Chem.* **4**, 114 (1974).

196. a) Hawker, P.N.; Kündig, E.P.; Timms, P.L. *J. Chem. Soc., Chem. Commun.* 730 (1978); (b) Hawker, P.N.; Timms, P.L. *J. Chem. Soc., Dalton Trans.* 1123 (1983).

197. Klabunde, K.J.; Low, J.Y.F. *J. Am. Chem. Soc.* 7674 (1974).

198. Lin, S.T.; Klabunde, K.J. *Inorg. Chem.* **24**, 1961 (1985).

199. Kündig, E.P.; Timms, P.L. *J. Chem. Soc., Chem. Commun.* 912 (1977).

200. Ozin, G.A.; McIntosh, D.F.; Power, W.J.; Messmer, R.P. *Inorg. Chem.* **20**(6), 1782 (1981).

201. Kasai, P.N. *J. Phys. Chem.* **86**, 4092 (1982).

202. Histed, M.; Howard, J.A.; Jones, R.; Tomietto, M. *J. Chem. Soc., Perkin Trans. 2*, **2**, 267 (1993).

203. Greaves, E.O.; Lock, C.L.; Maitlis, P.M. *Can. J. Chem.* **46**(24), 3879 (1968).

204. Fournier, R. *Int. J. Quantum Chem.* **52**(4), 973 (1994).

205. Zoellner, R.W.; Klabunde, K.J. *Inorg. Chem.* **23**, 3241 (1984).

206. Klabunde, K.J.; Murdock, T.O. *J. Org. Chem.* **44**, 3901 (1979) and references therein.

207. Anderson, B.B. *Diss. Abstr. Int. B.* **41**(11), 4108 (1981).

208. Klabunde, K.J.; Key, M.S.; Low, J.Y.F. *J. Am. Chem. Soc.* **94**, 999 (1972).

209. Bierschenk, T.R. *Diss. Abstr. Int. B.* **43**(7), 2202 (1983).

210. Bierschenk, T.R.; Juhlke, T.J.; Lagow, R.J. *J. Am. Chem. Soc.* **103**(24), 7340 (1981).

211. Firsich, D.W.; Lagow, R.W. *J. Chem. Soc., Chem. Commun.* **24**, 1283 (1981).

212. Klabunde, K.J. *J. Fluorine Chem.* **7**, 95 (1976).

213. Tanaka, Y.; Davis, S.C.; Klabunde, K.J. *J. Am. Chem. Soc.* **104**, 1013 (1982).

214. Ozin, G.A.; Power, W.J. *Inorg. Chem.* **16**(11), 2864 (1977).

215. Piper, M.J.; Timms, P.L. *J. Chem. Soc., Chem. Commun.* 50 (1972).

216. Francis, C.G.; Huber, H.; Ozin, G.A. *J. Am. Chem. Soc.* **101**(21), 6250 (1979).

217. Francis, C.G.; Huber, H.; Ozin, G.A. *Inorg. Chem.* **19**(2), 219 (1980).

218. Colin, G.F.; Huber, H.X.; Ozin, G.A. *Angew. Chem. Int. Ed. Engl.* **19**(5), 402 (1980).

219. Andrews, M.P.; Ozin, G.A. *Inorg. Chem.* **25**, 2587 (1986).

220. Colin, G.F.; Timms, P.L. *J. Chem. Soc., Dalton Trans.* **14**, 1401 (1980).

221. Wu, S.H.; Zhu, C.Y.; Huang, W.P. *Chin. Chem. Lett.* **7**(7), 679 (1996).

222. *Organometallic Polymers* (Eds: Karraer, C.; Shits, J.; Pimetten, S.). Moscow: Mir (1981).

223. a) Pomogailo, A.D.; Sevastianov, V.S. *Usp. Khim.* **52**(10), 1698 (1983); b) Pomogailo, A.D.; Sevastianov, V.S. *J. Macromol. Chem. Phys.* **25**, 375 (1985).

224. Pomogailo, A.D.; Uflyand, I.E. *Macromolecular Metal Chelates.* Moscow: Khimiya (1991).

225. Sahni, S.K.; Reedijk, I. *Coord. Chem. Rev.* **59**(1), 1 (1984).

226. Novikov, Yu.N.; Volpin, M.E. *Zh. Vses. Khim. Obshch. D.I.Mendeleeva.* **32**(1), 69 (1987).

227. Francis, C.G.; Timms, P.L. *J. Chem. Soc., Dalton Trans.* 1401 (1980).

228. Francis, C.G.; Timms, P.L. *J. Chem. Soc., Chem. Commun.* 466 (1977).

229. Francis, C.G.; Huber, H.X.; Ozin, G.A. *Angew. Chem. Int. Ed. Engl.* **19**(5), 402 (1980).

230. Ozin, G.A.; Andrews, M.P.; West, R. *Inorg. Chem.* **25**, 580 (1986).

231. Ryukhin, Yu.A.; Fokin, G.A. In: *Application of Organometallic Compounds in the Synthesis of Inorganic Films and Materials.* Moscow: Nauka (1983), p. 195.

232. Kuzharov, A.S.; Kutkov, A.A.; Suchkov, V.V. *Koord. Khim.* **6**(7), 1123 (1980).

233. Groshens, T.G.; Heine, B.; Bartak, D.; Klabunde, K.J. *Inorg. Chem.* **20**(11), 3629 (1981).

234. Morris, M.J. *Comprehensive Organometallic Chemistry*, Vol. 5 (Eds: Labinder, J.A.; Winter, M.J.). Oxford: Pergamon Press (1995).

235. Elschenbroich, C.; Nowothny, M.; Behrend, A. et al. *Angew. Chem. Int. Ed. Engl.* **31**(10), 1343 (1992).

236. *Comprehensive Coordination Chemistry* (Ed.: Wilkinson, G.). Oxford: Pergamon Press (1987).

237. Garnovskii, A.D.; Nivorozhkin, A.L.; Minkin, V.I. *Coord. Chem. Rev.* **126**(1–2), 1 (1993).

238. Garnovskii, A.D. *Russ. J. Coord. Chem.* **20**(5), 368 (1993).

239. Garnovskii, A.D.; Garnovskii, D.A.; Burlov, A.S.; Vasilchenko, I.S. *Ross. Khim. Zh.* **40**(4–5), 19 (1996).

240. Klabunde, K.J.; Li, Y.X.; Tan, B.I. *Chem. Mater.* **3**, 30 (1991).

241. Zenneck, U. *Chem. Unserer Zeit* **27**(4), 208 (1993).

242. Almond, M.J. *Annu. Rep. Prog. Chem., Sect. C: Phys. Chem.* **93**, 3 (1997).

243. *Active Metals: Preparation, Characterization, Applications* (Ed.: Fürstner, A.). Weinheim: VCH–Wiley (1996).

244. Freiser, B.S. *J. Mass Spectrom.* **31**(7), 703 (1996).

245. Chenier, J.H.B.; Howard, J.A,; Joly H.A.; LeDuc, M.; Mille, B. *J. Chem. Soc., Faraday Trans.* **89**(19), 3321 (1990).

Chapter 2

Direct Electrosynthesis of Metal Complexes

2.1 INTRODUCTION

Among the different types of direct syntheses [1–3] (electrosynthesis, metal-vapor synthesis of coordination and organometallic compounds [4], oxidative dissolution of metals in nonaqueous media [5] and mechano-synthesis [1]), electrosynthesis is not only the oldest but also the best method to obtain complex compounds starting from bulk zero-valent metals. As early as in 1882 Gerdes elaborated the electrochemical method for the synthesis of platinum(IV) hexaaminates, based on the anodic dis-solution of a platinum electrode in a solution of ammonium carbonate [6]. Later, in 1906, Scillard used this method for obtaining metal alkoxides [7]. The work mentioned had an episodic character. The systematic study of the use of electrosynthesis in coordination chemistry started in 1908 [8] with Chugaev, who synthesized the classic series of Verner's complexes ([Co(NH$_3$)$_6$]Cl$_3$, [Co(NH$_3$)$_5$Cl]SO$_4$ and [Co(NH$_3$)$_4$(NO$_2$)$_2$]NO$_3$) by electro-lysis with a cobalt anode and a platinum cathode; the chelate compounds (nickel(II) glyoximates) were also obtained using a nickel anode and dimethyl (or methylethyl) glyoximes. Nine decades after Gerdes' pioneering experiments, Lehmkuhl [9] in Germany and Garnovskii [10–13] in Russia "rediscovered" "direct electrosynthesis"; soon thereafter, Tuck [14] in Canada and Sousa [15] in Spain gave it additional impulse by developing simple electrochemical methods employing relatively unsophisticated appa-ratus, nonaqueous solvents, and sacrificial metallic anodes and cathodes. Various coordination and organometallic compounds have been obtained by this route, as the materials summarized in the reviews [2,3,14,16–20] and monographs [1,21–24] testify.

Electrochemical reactions do not require oxidants or reductants: only electrons ("universal chemical reagents" [17]) are used; they are the least expensive reagents (US$ $0.006\,mol^{-1}$, according to the Electrosynthesis Company Inc., New York, in 1998). Electrosynthesis can be made highly selective through control of electrode potentials, although in most instances a potentiostat is not required. Other advantages of this method are high purity of the products, high yields and selectivity, and the possibility of synthesis of complexes which are difficult to obtain by conventional methods. The characteristics of this synthetic route have been demonstrated for hundreds of reactions, and many industrial processes are nowadays practiced commercially worldwide.

The published material, an account of which is given in this chapter, reveals distinctly that electrosynthesis is highly reliable as a one-step method for obtaining metal complexes starting from zero-valent metals, and it should be used more frequently by chemists.

2.2 GENERAL CONCEPTS OF ELECTROCHEMICAL SYNTHESIS

2.2.1 Practical aspects

Electrosynthesis is the optimal method for carrying out redox reactions, because it works at normal temperature and has the following advantages:

- The addition of redox species to the reaction mixture is unnecessary.
- Compounds are produced by metal dissolution in soft conditions and with very simple equipment, independently of the metal being anode or cathode.
- It is possible to produce compounds which are very difficult to obtain by the classical route, especially those with the lowest metal oxidation state. This product selectivity, especially in organic electrosynthesis, demonstrates the power of electrochemistry.
- The electrosynthesized compounds are sometimes more reactive than those obtained by conventional methods.
- The working conditions required to obtain the electrodic reactions permit minimum contamination as a consequence of the low amounts of gases liberated, and avoid the danger of explosion during the electrolytic process. Frequently, the electrochemical methods are carried out under

milder conditions and at lower temperatures, leading to fewer reaction byproducts.

- At present prices of metals are lower than those of their compounds.

The use of a metallic salt to produce a coordination compound implies the presence of anion forming the salt. When the process is carried out in the conventional way (from metal salts MX and organic ligands HL), it is possible to obtain the complexes MLX instead of desirable product ML_n. The salt anion can participate in the formation of undesirable products.

Nevertheless, the electrolytical method makes it possible to obtain compounds with a higher metal oxidation state whenever the substitution of the sacrificial anode by a platinum electrode is possible.

Among the practical aspects, it is very important to recognize the different possibilities for carrying out any transformation and the existence of several starting materials in order to obtain the desired compound. Thus, economic factors should be taken into account before selecting the appropriate synthesis, and electrolysis should be applied only if it is the most economical method.

The application of direct electrosynthesis at an industrial level requires consideration of many factors, the most important of which are the following:

(1) The availability and price of the starting material.
(2) The quantity of product obtained (this is particularly important when the starting material is very expensive).
(3) The type and quantity of secondary products (they increase the price of obtaining the pure product). The presence of impurities can prevent use of the product in the case of, for example, the pharmaceutical industry.
(4) The price of the product isolated from the electrolytic medium.
(5) The maximum cell current, because this is one of the main factors which determine the quantity of the product obtained and thus the number of cells that should be used in the process. At the same time, the cell current depends on the total effective electrode area and on the current density. The latter parameter will be determined, for those systems having a single phase, by the substrate solubility in the mass-transport conditions. For this reason, as a useful practical rule we must take into account the necessity to have an electroactive species whose saturated solution will contain between 1 and 10% in order to obtain a current density around 100 mA cm^{-1}.

(6) The chemical and electrochemical stability of the electrolytic medium.
(7) The cell price (to avoid, if possible, the use of a membrane; use of cheaper electrodes and a high current density is recommended).

The relative importance of these and other factors depends on the nature of the process and especially on the application level.

In electrolysis processes it is very important to take into account the following experimental parameters:

(1) The electrode potential.
(2) The electrode material.
(3) The solvent and supporting electrolyte.
(4) The concentration of the electroactive species.
(5) The pH and concentration of all species that can react with the intermediates.
(6) The temperature and pressure.
(7) The mass-transport regime, which influences the maximum current density, the intermediate product velocities and the amount of the mixture between the reaction layer and the bulk solution. The mass-transport regime is determined by the electrolytic flow velocity and the movement of the electrodes.
(8) The geometrical form of the electrodes and the presence or absence of separators or membranes.

2.2.2 Solvent and supporting electrolyte

The electrosynthesis of organic compounds takes place by dissolution of the solute in an appropriate organic solvent (or a mixture of these) as a consequence of the low solubility of most organic compounds in water and also of the reduced potential range of this solvent.

In order to select the solvent it is necessary to take into account the following aspects:

- The solubility of the electroactive species and supporting electrolyte.
- Poor (or no) reactivity to the products.
- Facility of purification.
- Stability to the applied potential gradient of approximately $20\,V\,cm^{-1}$.
- Dielectric constant values.
- Low viscosity (especially if fast transport to electrodes is necessary).

- Adequate volatility to facilitate its elimination.
- Easy separation by precipitation of the product from the solvent.

The following standard solvents are the most widely used in electrosynthesis: alcohols, tetrahydrofuran (THF), dimethylformamide (DMF), dimethyl sulfoxide (DMSO), acetonitrile (AN), pyridine (Py); such exotic solvents as propylene carbonate, sulfolane, diglyme and hexamethylphosphortriamide [9,14,21] are also used. Among them (Table 2.1, below), acetonitrile and dimethylformamide are the most common. AN has the highest potential ranges and is especially useful because of its great resistance to oxidation and its low reactivity to ionic radicals. DMF is easier to oxidize than AN and for this reason it is not useful for anodic reaction; it is very resistant to reduction and reacts very weakly with anionic radicals [21].

The electrolysis can be carried out in the presence of emulsions or suspensions if adequate agitation or simultaneous ultrasonic treatment [25] is applied. With the purpose of decreasing the solution resistance in nonaqueous medium with a low dielectric constant, a high concentration of supporting electrolyte is added to the solute. However, this concentration does not insure a high conductivity, because of the tendency of many electrolytes to produce an ion-pair in most organic solvents.

The potential ranges in which several solvent-supporting electrolyte systems are electrochemically inert are reported in the literature. Nevertheless, it is recommended that the practical potential ranges in laboratory experiments should be determined. In organic media the selection of the supporting electrolyte is limited to halides, nitrates and perchlorates by solubililty considerations. The halides are not recommended, because of their ease of oxidation. The nitrates can be used for low anodic potentials, but for extremely high potentials tosylate, tetrafluoroborate and hexafluorophosphate are the best choices.

On the other hand, the cations most frequently used are those of aliphatic quaternary ammonium, because of their resistance to oxidation in the potential range of either one of the previously mentioned solvents. Salts of alkaline metals, which do not oxidize and are soluble in many solvents, are used as well.

Some important data concerning solvents and supporting electrolytes commonly used in direct electrosynthesis are reported in Tables 2.1 and 2.2.

TABLE 2.1
Properties of some solvents used in electrosynthesis [21a]

Solvent	m.p.,°C	b.p.,°C	ε^a	$E_{app.}$ V (vs SCE) $Pt_{(-)}$ $Hg_{(-)}$ $Pt_{(+)}$
AN	−45	+82	38	-2.7^c -2.8^d $+3.5^g$
DME	−58	+82	3.5	-2.95^f
DMF	−61	+153	38	-2.8^d -2.95^f
DMSO	+18	+189	47	-2.3 -2.8^d
Diglyme	—	+159	—	-2.95^f
HMP	+7	230–232 (at 739 mm Hg)	30	-3.3^c -2.95 $+1.0^c$
Pyridine	−42	+115	12.3	-2.2 -2.4 $+3.3^d$
Propylene carbonate	−49	+242	64	-2.2 -2.4 $+3.3^d$
Sulfolane	+28	+285	44	
THF	−80	+66	7.6	-3.2^c -2.95^f $+1.6^c$

^{a}Dielectric constant.
Supporting electrolytes: ^{b}1,2-dimethoxyethane; ^{c}LiClO$_4$; ^{d}[Et$_4$N][ClO$_4$]; ^{e}[Bu$_4$N][ClO$_4$];
^{f}[Bu$_4$N]Br; ^{g}[Et$_4$N]Br.
DME, dimethoxyethane; HMP, hexamethylphosphortrimide; the other abbreviations are
defined in the text.

TABLE 2.2
Limiting potentials for supporting electrolytes in THF and DMSO, V (vs. SCE)

Electrolyte	$Hg_{(-)}$/THF	$Pt_{(-)}$/DMSO	$Pt_{(+)}$/DMSO
[Bu$_4$N]Br	−2.95	−2.4	+1.45
[Bu$_4$N][ClO$_4$]	−2.94	−2.4	+2.10
[Bu$_4$N][BF$_4$]	−2.87		
[Et$_4$N]Br		−2.3	
LiCl		−2.68	+1.52
NaBF$_4$	−1.7		
LiClO$_4$	−1.95	−2.60	+2.10

2.2.3 Types of cells

Electrolytic cells can be classified as undivided and divided ones. In the former all the electrodes are located in the same compartment, while the divided cell has two compartments separated by a diaphragm (a ceramic one, an ion exchange membrane, etc.). The undivided cell which is generally used for electrosynthesis procedures in laboratory scale is shown in Fig. 2.1.

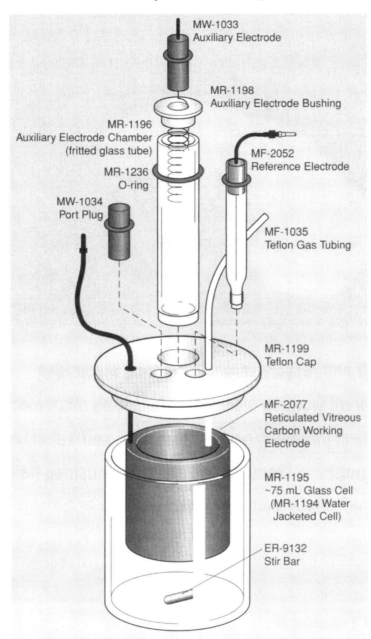

Fig. 2.1. Undivided electrochemical cell for electrosynthesis. (From *Bioanalytical Systems, A Handbook of Electroanalytical Products,* Bioanalytical Systems, Inc. with permission)

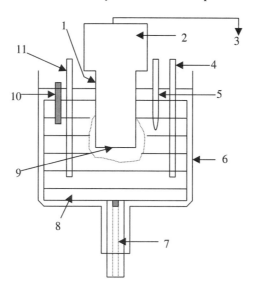

Fig. 2.2. Sonoelectrochemical cell used for electrosynthesis and voltammetric studies [25e]: 1, sonic horn; 2, transducer; 3, to sonic-horn control unit; 4, graphite counter electrode; 5, argon inlet for degassing; 6, Pyrex reservoir; 7, platinum-disk macro- or micro-electrode; 8, copper cooling coil connected to the thermostatted water-bath; 9, titanium tip; 10, platinum resistance thermocouple; 11, SCE reference.

Undivided cells have less resistance than divided ones; they are used when the reactant (or the product obtained) in the working electrode can suffer a transformation to an undesirable substance in the counter electrode, or when the reaction product in this electrode can react with the former product in the working electrode (the counter electrode can have a significant influence on the reaction and it can even change the nature of the product).

The most common cell for electrosynthesis processes is a 100-cm^3 tall beaker. The experiments can be run in a nitrogen atmosphere (or in an inert gas) to study the effect of oxygen on the process. For electrochemical synthesis with simultaneous ultrasonic treatment, it is possible to use the sonoelectrochemical cell reported in [25e] (Fig. 2.2).

2.2.4 Electrolytic parameters

Electrosynthesis experiments are normally carried out using a current range of 20–50 mA; the voltage required to obtain this current typically ranges from 10 to 50 V, depending on the electrolyte composition and the cell parameters. The voltage is determined not only by the electrode potential necessary for the reaction, but mainly by the electromotive force required to move the charge through a medium having a low dielectric constant. The resulting anodic current density is in the approximate range 10–30 mA cm^{-2}.

It should be emphasized that the above-mentioned parameter values are the most commonly found in the literature, although different values are possible.

Finally, it is important to point out that the most critical factor is the current density. An excessively low current density limits the quantity of product to be obtained, whereas a high current density produces an excess of heat in the cell.

The electrosynthesis of coordination compounds is carried out by using a sacrificial anode or cathode (which act as sources of metal ions to form complexes). Reactions with participation of the sacrificial anode are very frequently found in the literature [17,21,24]. The essence of electrosynthesis of complex compounds is described by the simple scheme (2.1), where Lig is a ligand [9].

$$M^+ + L + e^- \rightarrow M \cdots Lig \tag{2.1}$$

For reactions under solvation conditions, the electrosynthesis scheme could be described by equation (2.2), where L^0 is a ligand or its precursor [26].

$$M \underset{Ox}{\overset{-me^-}{\rightarrow}} [M(Sol)_n]^{m+} \underset{+L^0}{\overset{+me^-}{\rightarrow}} ML_x^0 \tag{2.2}$$

If ligands (LH) with an acidic EH group (E = NR, O, S, Se) take part in the process of electrosynthesis, then reactions (2.3) take place [2].

$$M - ne^- \rightarrow M^{n+}, \quad LH \Leftrightarrow L^- + H^+, \quad M^{n+} + nL^- \rightarrow ML_n \tag{2.3}$$

In the latter case, the mechanism proposes that anodic dissolution of the metal is a previous step to further complex formation (2.4) [17,21]:

$$
\begin{aligned}
\text{Anode}: \quad & nL^- + M \rightarrow ML_n + ne^- \\
\text{Cathode}: \quad & HL + ne^- \rightarrow nL^- + \frac{n}{2} H_2
\end{aligned}
\tag{2.4}
$$

For this reason, electrochemical reactions do not require the application of oxidants or reductants. This fact makes them attractive [21].

Other important advantages of electrosynthesis of coordination compounds, in comparison with the usual synthetic methods, are the ease of regulation, the reduced difficulty of obtaining compounds that are otherwise hard to access, and the possibility of conducting processes in much softer conditions with high yields [2,21,24].

2.3 ELECTROSYNTHESIS AT SACRIFICIAL ANODES

Metal complexes obtained under conditions of electrosynthesis include all types of coordination compounds: molecular adducts, metal chelates, π-complexes, di- and trinuclear homo- and heterometallic structures, as well as some σ-organometallic compounds. However, the majority of electrochemically synthesized complexes belong to the first two types mentioned above.

2.3.1 Electrosynthesis of molecular complexes (adducts)

Molecular complexes obtained by electrochemical methods are reported for various adducts of organic N,O,P-donors with metal halides (pseudohalides) and organohalides [14,17,19,21,24,27–31]. The first ones have the general formula $MX_n \cdot pL$, while for the second group the formula is $R_mMX_{n-m} \cdot pL$, where M are the metals of groups I (Cu, Ag, Au) [14,17,19, 21,24,32–37], II (Mg, Zn, Cd, Hg) [14,17,21,24,31,36–41], III (In, Ga, Tl) [14,17,20,21,24,32,36,37,42–46], IV (Sn, Pb, Ti, Zr, Hf) [14,17,21,24,27, 35,37,47–49], V (V, Nb, Ta) [14,17,21,44], VI (Cr) [14,28,32,44], VII (Mn) [14,17,21,24,39] and VIII (Fe, Co, Ni, Pd) [14,24,30,32,50–52], and the acthinides Th [17,21,27,53] and U [28].

The following compounds are used as ligands in the complexes examined: N-donors as amines (am) [24,32,33,36,37], acetonitrile (AN) [14,17,21,24,27 ,35,39,41,43,44,46,47,50,53], pyridine (Py) [14,43], bipyridyl (bipy) [14,24,27, 32,35,37,39,45,48,52,54], o-phenanthroline (phen) [14,24,27,32,35,37,39,45, 46,48,52,54] and imidazole [14]; P-donors as phosphines [14,33,34,36,37, 43,45,51] and diphosphines [14,32,36,39]; O-donors as methanol [39], diethyl ether [34], dioxane [14,24], DMSO [14,24,27,32,42,44,53], DMF [28] and triphenylphosphine oxide [27,54]; and S-donors as carbon disulfide [36]. Compounds used as R substituents are as follows: Alk, Ar [14,24],

X = OH [31], Hal [14,17,21,24,28,33,39,41,43,47,50], CN [14], NCO [29] and NO$_3$ [54].

Molecular complexes are usually obtained under conditions of electro-synthesis by anodic dissolution of a metal in the presence of ligands and adduct-forming compounds (bipy, phen, phosphines etc.) or donor solvents (AN, Py, DMSO, DMF, etc.). At the same time, in some cases, the electro-chemical synthesis could be accompanied by original transformations. Thus, by the dissolution of cadmium in acetonitrile solution containing *N*-(hydro-xyethyl)salicylidenimine and phenanthroline, instead of an azomethine che-late [55], the adduct Cd(OH)$_2$(phen)$_2$ was obtained [31]. The structure of this complex is confirmed by X-ray diffraction: the compound has an octahedral structure with a CdN$_4$O$_2$ coordination kernel [31]. Ionic complexes of either cationic [14,17,24,27,29,30,32,42] or anionic [14,17,21,24,27,28,33,41,43,47, 51,54] types could be attributed to the molecular coordination compounds. Cationic complexes are present in a very wide range of compounds [14,24], synthesized by means of the anodic dissolution of metals in organic solvents containing the mineral acid HBF$_4$ or HClO$_4$. The ammonium salts of HClO$_4$ also serve as sources of perchlorate ions [29,30].

The series of cationic complexes with the general formula [ML$_n$][BF$_4$], where L = AN or DMSO; $n = 4$ or 6 and $m = 2$ or 3, were obtained by electrolytic dissolution of metals (Zn, Cd, In, Ti, V, Cr, Co, Ni, Fe) in acetonitrile solution or DMSO in the presence of HBF$_4$ [14,24]. The electro-chemical dissolution of the copper anode in acetonitrile in the presence of 2,2′-bipy, (1*H*)pyrid-2-one and tetramethylammonium perchlorate leads to copper(II) bis(2,2-bipy)isocyanatoperchlorate [Cu(C$_{10}$H$_8$N$_2$)$_2$(NCO)]ClO$_4$ [29]. According to the X-ray diffraction data [29], the cation this complex represents a square pyramid with the cyano group at the top, which is coordinated with copper through a nitrogen atom; this represents an inter-esting topic for discussion—the ambidentate binding of pseudohalide ions [56–58]. The electrosynthesis process with a cobalt anode under the same conditions, but in the presence of 1,10-phenanthroline, promotes the forma-tion of a complex which, according to the X-ray diffraction study, has an octahedral structure at the expense of bidentate binding of the phenanthro-line ligand (N,N-coordination) and a carbonate ion (O,O-coordination) [30].

Amongst the anionic complexes, compounds of the type [Et$_4$N]$_m$[MBr$_4$] are well described in the literature [14,24]. They have been obtained by electrosynthesis with the participation of zero-valent metals, bromine and tetraethylammonium bromide in a benzene–methanol solution with Au, Cd, In, Sn, Ti, Hf, Co, Ni, and Fe as metals.

The adducts of thiolates and selenolates, obtained by electrosynthesis, with N- [35,37,45,46,49,52] and P-donors [37,45,46,49], could be attributed to molecular complexes of a neutral type. The general method for obtaining those metal thiolates and selenolates is based on the electrochemical cleavage of the X–X (X = S, Se) bond (reaction (2.5)) [37,40,45,49].

$$n/2\,Ar\!-\!X\!-\!X\!-\!Ar \xrightarrow{M} M(XAr)_n \qquad (2.5)$$
$$\mathbf{I}$$

Adducts or the type $M(XA)_n \cdot L_m$ are formed in the presence of the ligands shown above. The same transformation (2.5) should perhaps be characteristic for ditellurides also. At the same time, it is necessary to mention that metal thiolates and their adducts could also be obtained by starting from thioalcohols [17,21,35,46,49,52,59] (Table 2.3, below). In this respect, the electrosynthesis of metal alkoxides [7,17,21,60,61] and their subsequent use as precursors for sol–gel processes could have practical interest.

The structures of the type **I** complexes and their adducts have been studied by various physical methods, in particular X-ray diffraction [35,49,55]. On the basis of these data, it was established that the adduct of the complex of o-toluidine sulfide with o-phenanthroline and acetonitrile corresponds with the dimer, $[Cu(SC_6H_4CH_3\text{-}o)(C_{12}H_8N_2)]_2$, with sulfide bridges connecting the copper atoms [35]. Its magnetic properties, which unfortunately have not yet been studied, are of especial interest: this complex should be antiferromagnetic [62]. There is an octahedral environment due to the metal in the complex monomers $[Sn(SC_6H_5)_4 \cdot bipy]$ [49] and $[Co(SC_6H_5)_2 \cdot phen]_2(ClO_4)$ [52].

The series of complexes with the general formula $M(SR)_2L_2$ (where M = Co, Ni; R = Ph, 2,3,5,6-tetrafluorophenyl, $C_6H_4Me\text{-}o$, $2\text{-}C_{10}H_7$; L = phen, bipy [52]), have been obtained by galvanostatic electrolysis in acetone or acetonitrile. Measurement of their magnetic properties (for the cobalt complex $[Co(SPh)_2(phen)_2]$, $\mu_{eff.} = 2.70\ \mu_B$, and for the nickel analogue 3.30 μ_B) and the composition testify to the octahedral configuration. The adducts with the general formula $CuSR \cdot nL$ (where R = Ph, $o,m,p\text{-}MeC_6H_4$, 2,3,5,6-tetrafluorophenyl; L = $Me_2NCH_2CH_2NMe_2$, PhNCS, $t\text{-}BuNCS$, PPh_3; n = 0.5, 1.5, 2) were synthesized [49] by the electrochemical dissolution of copper in acetonitrile solution of thiols and ligands (L).

TABLE 2.3
Electrosynthesis of metal adducts, alkoxides and related compounds

Metal	Initial system	Products	Ref.
Al	NH_3 (liquid), NH_4Br	$Al(NH_3)_6Br_3$ + $[Al(NH_2)(NH)]_n$	63a
Pd	HX (X = Cl, Br)	PdX_2	63b
Sb, Fe, Ti, Hg, Re, Ni, Mn	ROH	$M(OR)_n$ or $MO(OR)_n$ $Re_4O_2(OMe)_{16}$	14a, 64–66
In, Sn	CH_3CN, $R(OH)_2$ (R = 1,2-dihydroxybenzene, 2,3-hydroxynaphthalene, 2,2'-dihydroxybiphenyl, 1,2-dihydroxytetrabromobenzene)	$In[OR(OH)]$ $Sn(O_2R)$	67,68
Mn, Cd, Fe, Ni, Cu, Mo	$EtOH + H_2O$, NH_4SCN; Me_4NCl or Me_4SCN	$(Me_4N)_n[M_m(SCN)_l]$, $n = 2$ or 4; $m = 1$ or 2; $l = 4$ or 6	69
Zn	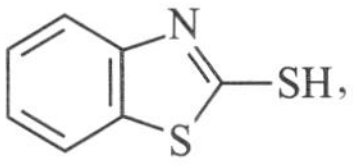 —SH, bipy	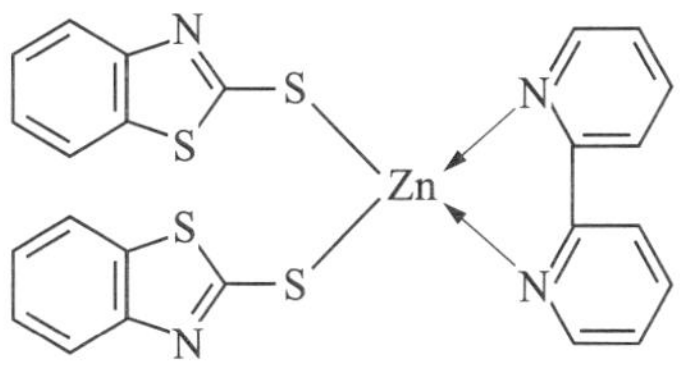	70
Cd, Cu, Tl, Sn, Zn	CH_3CN, toluene, Ph_2Se_2	$M(SePh)_2$ (M = Cd, Zn) MSePh (M = Cu, Ag, Tl) $Sn(SePh)_4$	37, 45
	In the presence of phen	$Cd(SePh)_2 \cdot 2phen$ $Cu(SePh)_2 \cdot phen$	
	In the presence of PPh_3	$Cu(SePh)_2 \cdot 1.5PPh_3$	
Cd, Co, Fe, In, Ni, V, Zn, Ti, Mn	DMSO, HBF_4 CH_3CN, HBF_4	$[M(DMSO)_6(BF_4)_2]$ (M = Cd, Co, Mn, Ni, V, Zn) $[M(DMSO)_6(BF_4)_3]$ (M = In, Fe) $[M(CH_3CN)_6(BF_4)_3]$ (M = In, Ti)	44
Cr	DMSO, HBF_4 CH_3CN, HBF_4	$[Cr(DMSO)_6][BF_4]_3$ $[Cr(CH_3CN)_n][BF_4]_2$, $n = 4,6$	44
In	$HClO_4$, MeOH, DMSO	$[In(DMSO)_6][ClO_4]_3$	42
In	DMSO, benzene, Et_4NX (X = Cl, Br)	$InX_3 \cdot 3DMSO$	42

TABLE 2.3   (Contd.)

Metal	Initial system	Products	Ref.
Mn, Cd, Hg, Co, Tl, Cu, V, Ga	CH_3CN, $(EtO)_2P(S)SH$ (= HL) In the presence of phen	ML_2, ML_3, TlL $CoL_3 \cdot phen$, $MnL_2 \cdot phen$, $VL_2 \cdot phen$, $GaL \cdot phen$	70
Pd	HX (X = Cl, Br), further addition of MX {M = NH_4, $(CH_3)_4N$}, LH_2 (phen or bipy hydrates)	$M_2[PdX_4]$ $M_2[PdX_4] \cdot LH_2$	63b
Ti	n-C_3H_7NCl, Cl_2, $SOCl_2$ Et_4NBr; $Br_2 \; C_6H_6 = 1:5$ Et_4NBr; $Br_2 \; C_6H_6 = 1:3$	$(n\text{-}Pr)_4NTiCl_5$ $Et_4NTi^{III}Br_4$ $[Et_4N]_2Ti^{IV}Br_6$	47
Hf, In, Fe, Ni, Th, Ti, V, Zr	X_2, CH_3CN (X = Cl, Br)	$MX_4 \cdot 2CH_3CN$ (M = Ti, Zr, Hf), $InX_3 \cdot nCH_3CN$ (n = 2, 3), $VCl_2 \cdot 2CH_3CN$ $ThX_4 \cdot 4CH_3CN$, $VBr_2 \cdot CH_3CN$ $MX_2 \cdot CH_3CN$ (M = Fe, Ni)	27, 39, 47, 71, 39
V	I_2, CH_3CN	VI_2	
Zr, Hf	Cl_2, $SOCl_2$, n-Pr_4NCl Br_2, C_6H_6, Et_4NBr Ethylenediamine, benzene, methanol	$(n\text{-}Pr)_4NMCl_5$ $(Et_4N)_2MBr_6$ $MBr_4 \cdot en$ (M = Zr, Hf)	
U	Cl_2 or Br_2, CH_3CN, N_2 Cl_2 or Br_2, CH_3CH, O_2 DMF, R_4NX (X = Cl, Br), CH_3CN, N_2 DMF, R_4NX, CH_3CN, O_2 DMF, I_2, CH_3CN	$UCl_4 \cdot 4CH_3CN$ $UBr_4 \cdot 2CH_3CN$ UO_2X_2 $(R_4N)_2UX_6$ $(R_4N)_2UO_2X_4$ $UI_4 \cdot 4DMF$	28
In	X_2 (X = Cl, Br), benzene, CH_3OH or DMSO	$InX_3 \cdot nSolv$, n = 0 or 3	72, 43
In, Fe, Sn	In the presence of R_4NX (R_4 = Et_4, Et_3Ph)	$(R_4N)_2InX_5$ Et_4NFeBr_4, $(Et_4N)_2SnBr_6$	42, 72, 73
Mn	Br_2, CH_3OH, C_6H_6 Br_2, CH_3CN	$MnBr_2 \cdot MeOH$ $MnBr_2 \cdot nCH_3CN$	39 39

TABLE 2.3   (Contd.)

Metal	Initial system	Products	Ref.
Cd, Ni	Benzene, CH_3OH, Et_4NBr, Br_2	$(Et_4N)CdBr_3$ $(Et_4N)_2NiBr_4$	73
Cu	Br_2, C_6H_6, CH_3OH, Et_2O	$CuBr$ $CuBr_2 = 0.28{:}0.72$	39
In	C_6H_6, CH_3OH, R_4NCl, Cl_2 ($R_4 = Et_4$ or Et_3Ph)	$(R_4N)_2InCl_5$	42, 72
	Et_4NI, I_2, CH_3OH	$[Et_4N][InI_2]$	42, 74
Cu	CH_3CN, RX, Et_4NX ($R = Me$, Ar; $X = Cl$, Br, I) CH_3CN, PhBr, bipy CH_3CN, CH_3I, bipy	$[(C_2H_5)_4N]_n[Cu_2X_m]$ ($n = 1$–3; $m = 2$–5) $CuBr{\cdot}bipy$ $CuI{\cdot}bipy$	33
Tl	Nonaqueous solvent, SiO_3^{2-}, R_4NF, HF	TlF, Tl_2SiF_6, $Tl_3[SiF_6]F$	75
Ga	CH_3CN, HX, PPh_3 ($X = Cl$, Br, I)	$[Ph_3PH]_2[Ga_2X_6]$	20
Cd, In, Tl	α, ω-Alkanedithiols $HS(CH_2)_nSH$ ($n = 2$–6), CH_3CN In the presence of phen or bipy (L) $HS(CH_2)_nSH$ ($n = 3$–6), CH_3CN	$Cd[S_2(CH_2)_n]$ $In[SR(SH)]$, $Tl_2(S_2R)$ $Zn[S_2(CH_2)_n]$ $M[S_2(CH_2)_n]{\cdot}L$ ($M = Zn$, Cd) M_2S_2R ($M = Cu$, Ag)	76 77 76
Cu, Ag	In the presence of PPh_3 or dppm (L)	$Cu_2S_2R{\cdot}nL$ (not all combinations)	78
In	CH_3CN, (I_2), RSH ($R = Alk$, Ar) In the presence of bipy or phen (L)	$InI_n(SR)_m{\cdot}CH_3CN$ $In(SR)_n$ $InI_n(SR)_m{\cdot}L$	46
Cd, Co, Au, Ag, Zn, Cd	CH_3CN, PPh_2H In the presence of S and toluene	$M(PPh_2)_2$ ($M = Cd$, Co, Zn) $M(PPh_2)$ ($M = Ag$, Au) $Cd(S_2PPh_2)_2$	79
Cr, Co Mn, Fe, Ni	RCOOH, CH_3CN ($R = Et$, n-C_7H_{15}, Ph, Me)	$CrL_3{\cdot}2H_2O$ ($R = Et$) ML_2 ($R = Et$; $M = Co$, Ni) FeL_2 ($R = Et$, n-C_7H_{15}) CrL_3 ($R = Ph$) MnL_2 ($R = Me$, Et, Ph, n-C_7H_{15})	80a

TABLE 2.3   (Contd.)

Metal	Initial system	Products	Ref.
Fe, Co, Ni, Cu	Lactic acid (H_2LA), acetone	$M(HLA)_n \cdot mH_2O \cdot k(CH_3)_2CO$	80b
Cu	H_2O, HL; L = propionic, benzoic, oxalic, succinic, phthalic, isobutyric, salicylic, acetic acids; glycine, L-alanine, histidine, sodium pyruvate	$CuL_2 \cdot H_2O$	66
Cr, Cu, Fe, Mg, Ni, V, Zr, Mo, W, Sn	DMSO, SO_2	$Cr_2(DMSO)_{20}(S_2O_7)_3$ $[M(DMSO)_6][S_2O_7]$ (M = Cu, Mg, Zn) $Fe_2(DMSO)_{10}(S_2O_7)_3 \cdot SO_2$ $Ni_2(DMSO)_7(S_2O_7)$ $Zr_2(DMSO)_{10}(S_2O_7)_3$ $V_2(DMSO)_{11}(S_2O_7)_3$ $(Mo, Sn)(DMSO)_4(S_2O_7)$ $W_2(DMSO)_6(SO_4)_3$	81, 82
Cr, Co	H_2O + EtOH; Me_4NCl or Me_4NSCN In the presence of bipy	$(Me_4N)_3[Cr(SCN)_6]$ $(Me_4N)_2[Co(SCN)_2]$ $[Co(bipy)_2(SCN)_2]$	69
Co, Ni, Hg, Tl, Sn Pb	CH_3CN + RSH (R = Alk, Ar) PhSH In the presence of phen or bipy (L)	$M(SR)_2$ (M = Co, Ni, Sn, Hg) TlSR $Pb(SPh)_2$ $Ni(SR)_2 \cdot 2L$ $Sn(SR)_4 \cdot L$	46, 52, 83
Cu	RSH, PPh_3 (R = Ph, o, m, p-MePh etc.)	$CuSr \cdot nPPh_3$	84
Co	Br_2, CH_3CN Br_2, CH_3OH, C_6H_6 In the presence of Et_4NBr	$CoBr_2 \cdot 2CH_3CN$ $CoBr_2$ $(Et_4N)_2CoBr_4$	39 73
Au	Benzene, CH_3OH, Et_4NBr, Br_2	$Et_4N \cdot AuBr_4$	73
Au, Ag	PPh_3, CH_3CN, benzyl chloride PPh_3, CH_3CN, HCl	$AuCl(PPh_3)_2$ $AgCl(PPh_3)$ $AuCl(PPh_3)$	34
Co	CH_3CN, HCl, PPh_3	$[Ph_3PH]_2[CoCl_4]$	51

TABLE 2.3 (Contd.)

Metal	Initial system	Products	Ref.
Cu	2,2'-dipyridylamine (=HL), CH_3CN/N_2, dppm	$CuL \cdot 0.5dppm$	36
	2,2'-dipyridylamine (=HL), CH_3CN/N_2, dppe	$CuL \cdot 1.5dppe$	
Cu	PPh_3, CH_3CN, benzyl chloride	$CuCl \cdot PPh_3$	34
	PPh_3, CH_3CN, HCl	$CuCl \cdot PPh_3 + Cu_2Cl_2(PPh_3)_3$	
	PPh_3, CH_3CN, HX X = Br, I	$Cu_2X_2(PPh_3)_3$	
Ni	CH_3CN, PPh_2H	$Ni(PPh)_2 \cdot PPh_2H$	79
	CH_3CN, PPh_2H, PhSCN	$[Ni(PPh_2)(PhSCN)]_2$	
Cu	CH_3CN, dppm, $Cl_4C_6(1,2-OH)_2$	$Cu_2[OC_6Cl_4(OH)]_2 \cdot (dppm)_2$	78
Mn	CH_3CN pymtH or 4,6-Me_2pymtH, bipy or phen	$[Mn(pymt)_2 \cdot phen]$ $[Mn(4,6-Me_2pymt)_2 \cdot phen]$ $[Mn(4,6-Me_2pymt)_2 \cdot bipy]$	85
Th	Acetone, Hoxine	$Th(oxine)_4$	27
Sn	$AgClO_4$, CH_3CN	$Sn(ClO_4)_2 \cdot 2CH_3CN$	18
V	H_2O_2	After increasing the pH with KOH: $K[VO(O_2)_2(H_2O)]$ $K_3[VO(O_2)_3] \cdot 3H_2O$	18
Mo, W Nb, Ta, Ti, V	H_2O_2, HF	$[MO(O_2)F_4]^{2-}$ $[M(O_2)F_5]^{2-}$ $[VO(O_2)_2F]^{2-}$	18, 86
Fe	Indene, 1-methylindene, 1,3-dimethylindene, 3-phenylindene (HL)	FeL_2	87

en, ethylenediamine; dppm, bis(diphenylphosphino)methane; dppe, bis(diphenylphosphino)ethane; pymtH, pyrimidine-2-thione; 4,6-Me_2pymtH, 4,6-dimethylpyrimidine-2-thione; Hoxine, 8-hydroxyquinoline.

Thus, the electrosynthesis of molecular complexes is represented by a convincing number of products obtained, as well as by the metal chelates (see Section 2.3.2). A summary of the literature data on the electrosynthesis of adducts and related compounds are presented in Table 2.3.

Experimental Procedure

The experimental synthetic methods corresponding to some of the compounds reported in Table 2.3 are described below.

(a) Direct electrochemical synthesis of some copper(I) complexes [33]

As described in [33] the general type of cell used can be represented by the following scheme:

$$Pt_{(-)}/CH_3CN + RX + bipy/Cu_{(+)}$$

or

$$Pt_{(-)}/CH_3CN + RX + C_2H_5)_4NX/Cu_{(+)}$$

The reaction conditions in the case of $[(C_2H_5)_4N][Cu_2Cl_3]$ are, for example: $RX = C_6H_5CH_2Cl$; $15\,cm^3$ acentonitrile; $1.0\,g$ $(C_2H_5)_4NX$; $V = 5\,V$; $I = 20\,mA$; $t = 5\,h$; anode loss = 290 mg. Found (%): Cu, 30.1; H, 35.0. Calculated (%): Cu, 29.3; H, 34.9.

The anionic complexes were isolated by conventional routes and the reduction of the volume of the electrolytic solution by almost 50% gave the precipitation of a solid, which was washed and/or recrystallized and dried in vacuo. The yields were approximately of 80%, based on the quantity of copper dissolved.

(b) Electrochemical preparation of group IVA halide complexes MX [47]

Each metal was used as the anode of a cell containing a nonaqueous phase with a platinum wire as the auxiliary electrode.

For example, in case of the Ti/Br complex the experimental conditions were: solution composition (volume in cm^3) CH_3CN 50; Br_2 2. $V = 10$ V; $I = 200$ mA; $t = 4.5$ h; anode loss = 1.34 g. According with the analytical results the adduct formula was $TiBr_4 \cdot 2CH_3CN$.

The chemical efficiency of the process, expressed as moles of complex obtained per mole of metal dissolved, was generally in the range 80–95%. For this reason, no attempt was made to study the effect of electrolytic conditions on the overall chemical efficiency.

The electrochemical technique gave the possibility of obtaining the bis(acetonitrile) adducts of the tetrachlorides and bromides, except in those cases where gram quantities of tetraethylammonium halide were present, when the appropriate anionic halogen complexes were obtained. Thionyl chloride proved to be the most satisfactory solvent for the preparation of the anionic chloro complexes.

(c) Manganese ethylate [65b]

Ethylcellosolve (30 cm^3) was added to ethanol solution of NaBr (2.3%, 165 cm^3). The electrolysis (Mn anode and rotating steel cathode) was carried out at 55–60°C and at an anodic current density of 5.1–6.0 A dm^{-2} during 6.5 h (nitrogen atmosphere). Then the solution was placed (in an N$_2$ atmosphere) in the closed flask. After two days the electrolyte was decanted from the dense solid, to which absolute ethanol (150 cm^3) was added. After another two days this procedure was repeated. The last fraction of washing ethanol contained practically no NaBr or ethylcellosolve. The ethanol residues were removed by pumping at 25–30°C. The product reached constant weight after 12–13 h in vacuum. The product formed (56.2 g, 97.5%) was a gray crumbly powder.

(d) Synthesis of cobalt(II) butanolate [64]

A solution of 9 g of LiClO$_4$ and 0.75 g of LiCl in 150 cm^3 of butanol was electrolyzed between two cobalt electrodes at 25°C. The electrolytic parameters were: $I = 0.5$ A; $V = 23$–25 V; $Q = 9.3$ A h; anode loss = 100% of theoretical.

The suspension of the reaction product was filtered and washed with 230 cm^3 of butanol. After drying, 31.8 g of cobalt butanolate (90% of theoretical) was obtained. In the reaction with acetylacetone 80% of the theoretical yield of butanol was obtained. For C$_8$H$_{18}$CoO$_2$, Found (%): Co, 29.8. Calculated (%): Co, 28.7.

(e) Electrochemical preparation of bimetallic adducts [Cr$_2$(DMSO)$_{20}$][S$_2$O$_7$]$_3$ and [Fe$_2$(DMSO)$_{10}$][S$_2$O$_7$]$_3$·SO$_2$ [81]

See Example (a) under Experimental procedure in Section 2.3.4.

2.3.2 Electrosynthesis of metal chelates

Earlier, only some chelate-forming ligands, such as dimethylglyoxime [2,7,21,24], β-diketones [2,14,24], azomethines **II** and **III** [2,17,21,24], oxyphenylazoles and their benzannelated derivatives **IV** [2,24], 2-acetylpyrrole **V** [14], and 3-hydroxy-4-pyrone **VI** [14], were used for the reactions of electrosynthesis of metal chelates.

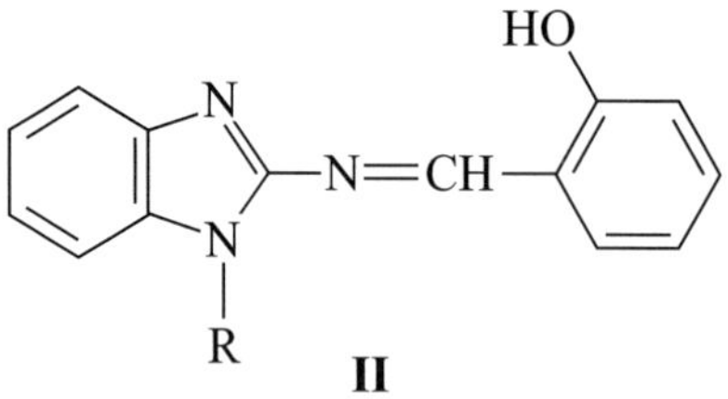

The electrosynthesis of chelates based on chelate-forming ligands with an acidic XH group where X = NR, OR, S will be discussed below; their usual synthesis is reported in [88,89]. Chelates with N,O,S-ligand systems incorporating various metal-containing rings have now been electrosynthesized. The greatest numbers of chelates with a six-membered metal ring, among those complexes obtained electrochemically, are reported for chelates of β-diketones [14,17,21,24,27,90–97] and azomethinic ligands [10,17,21,24,55,98–116]. Thus, electrochemical synthesis was used to prepare β-diketonates of the following metals: copper [14,24,91], cadmium and zinc [14,24,92], aluminum and indium [91,94], titanium, zirconium and hafnium [14,91,94], vanadium, niobium and tantalum [14,94], chromium and molybdenum [91,94], manganese [14,24,91], iron, cobalt and nickel [14,24,91,97], cerium [24], dysprosium [96], uranium [93,95] and thorium [27]. The structure of the adduct Cd(acac)$_2$·phen has been determined by X-ray diffraction [92].

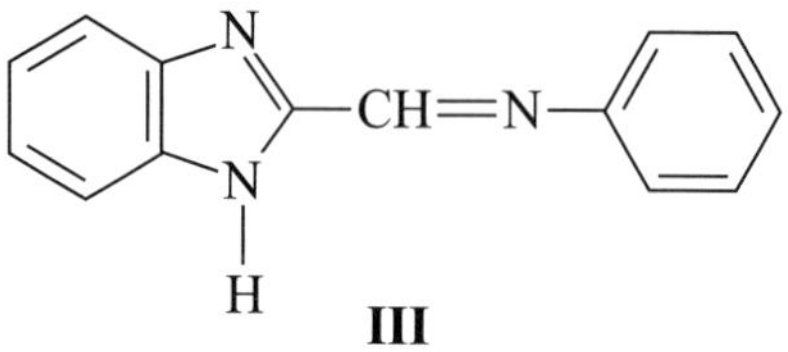

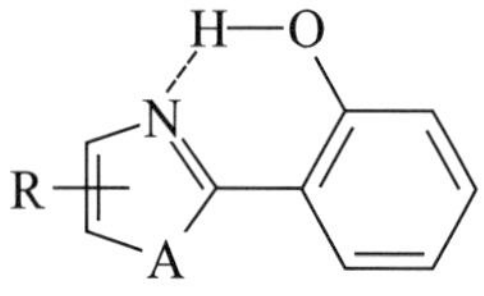

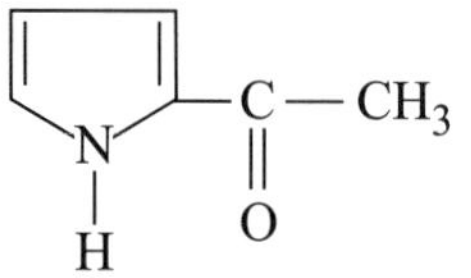

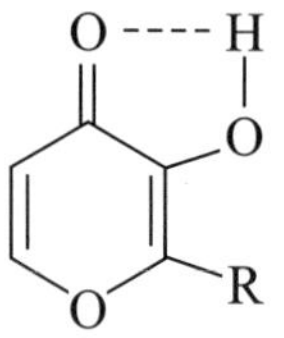

A systematic study [94] showed that β-diketonates with the compositions ML$_2$ (M = Ni), ML$_3$ (M = Al, In, Co, Fe, Ti, Zr), ML$_4$ (M = Zr, Hf), ML$_2$·solv (M = Ni, Fe), ML$_3$Cl (M = Ti, Zr), TiL$_2$(OMe) and ZrL$_3$OMe, where L is a diketone anion, are obtained in good yields on the basis of various β-diketones and their fluoro-substituted derivatives under the conditions of electrosynthesis. The syntheses of these chelates were carried out

TABLE 2.4
Conditions for the preparation of some metal acetylacetonates [94]

Compound	Solvent (cm^3)	Electrolyte (g)	Water (g)	Hacac (g)	E_F	Anode weight loss (g)	Yield (%)
Ni(acac)$_2$	MeOH (50)	NaCl (2)	60	40	0.44	3.18	93
Co(acac)$_2$	EtOH (50)	KCl (2)	60	40	0.51	6.54	63
Co(acac)$_2$	DME (100)	LiClO$_4$ (1.75) LiBr (2.5)	—	75.4	0.51	5.80	67
Co(acac)$_2$	Diglyme (100)	LiClO$_4$ (1.25) LiBr (2.5)	—	75.4	—	—	89
Ni(acac)$_2$	DME(100)	LiClO$_4$ (1.25) LiBr (2.5)	—	75.4	0.41	6.30	87
Ni(acac)$_2$	PDC (100)	LiClO$_4$ (1.25) LiBr (2.5)	—	75.4	—	—	80
Ni(acac)$_2$	Py (100)	LiClO$_4$ (1.25) LiBr (2.5)	—	75.4	—	—	68
Mn(acac)$_2$	THF (39.4)	LiClO$_4$ (0.95) LiCl (0.04)	—	43.3	0.58	5.88	86

by using liquid β-diketones or their 15–20% solutions in methanol, ethanol, AN and DMF, with LiCl, LiBr, NH$_4$Cl or Bu$_4$NBF$_4$ as supporting electrolytes (Table 2.4).

The electrochemical oxidation of uranium leads to chelates of the type UL$_4$ and UO$_2$L$_2$ (LH = diketone) [93,95]. In addition to these complexes, the compound having the composition UO$_2$L$_2$(HL)$_{0.5}$ was also isolated [95]; the structure VII was proposed from IR spectroscopy data and requires more complete proof.

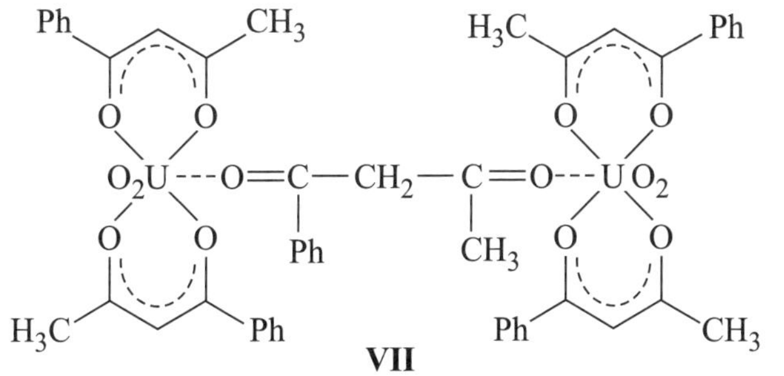

VII

The chelates of Schiff bases have been obtained by electrosynthesis in acetonitrile solutions of ligands by using a platinum electrode as cathode and an anode made from the corresponding metal. They are represented by metal chelates **VIII**, obtained on the basis of ligand systems with various substituents (R) in the amine fragment [55], as well as by the complexes of heterocyclic amines and azomethines [10,17,21,24,98].

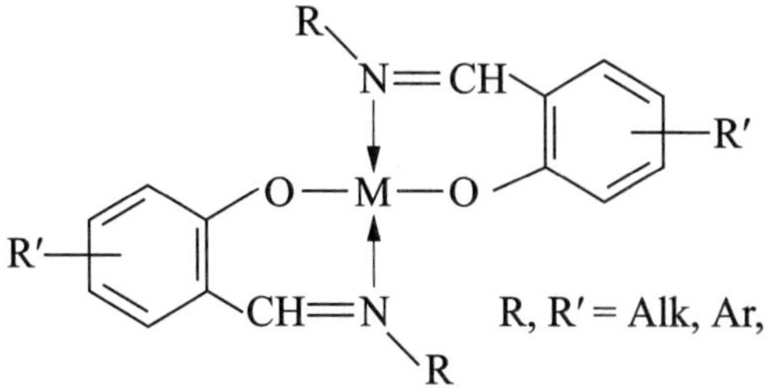

VIII

Besides complexes containing alkyl or aryl substituents [55] (R = *i*-Pr, Bu, *p*-MeC$_6$H$_4$, *p*-ClC$_6$H$_4$, *p*-NO$_2$C$_6$H$_4$) on the azomethinic nitrogen atom, the majority of *o*-hydroxyazomethines synthesized by electrochemical methods [99–115] have the coordinationally active fragments **VIII** (R = 2-Py [99,101]; R = CH$_2$CH$_2$Py-2 [100,102]; R = CH$_2$CH$_2$NMe$_2$ [100]), **IX** [105], **X** (X = S [106,108,111]; X = Se [110]), **XI** [104]; or **XII** [104].

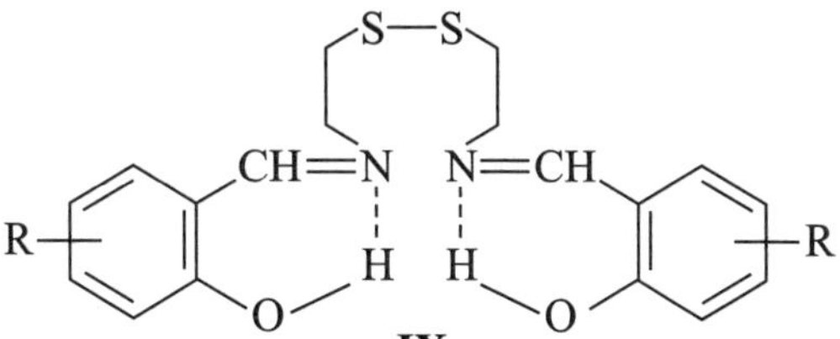

IX

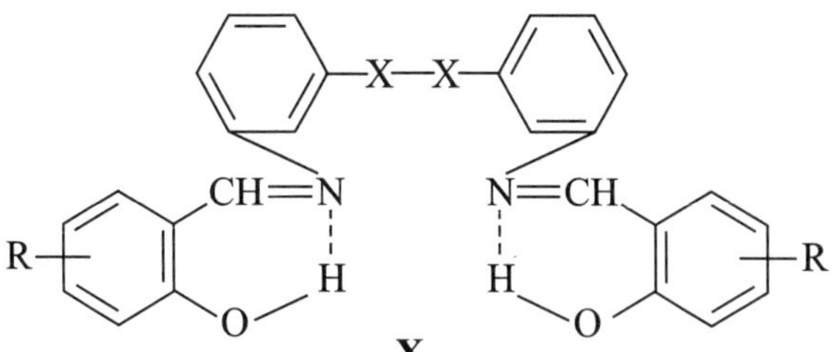

X = S, Se   **X**

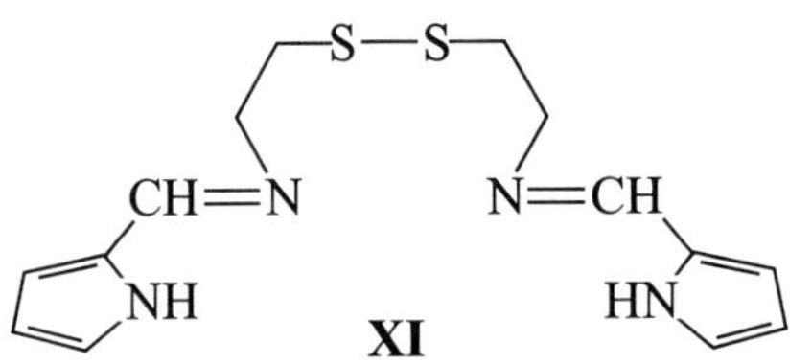

The analysis of X-ray diffraction data testifies [101] that the N-atom in the pyridine ring does not take part in the coordination, within the copper complex, of the ligand **VIII** (R = 2-Py-3-Me, R′ = 5-OMe), i.e. this is formally a tridentate ligand system behaving as a bidentate one; the metal-containing rings MN_2O_2, with the participation of a phenolic oxygen atom and an azomethine nitrogen atom, are found in the complex compounds described. The same situation is characteristic for other copper complexes **VIII** (R = CH_2CH_2Py, R′ = 4,6-OMe): they have a flat-square structure with a noncoordinated nitrogen atom on the pyridine ring [100]. At the same time, X-ray diffraction data indicate the participation of the pyridine ring in the coordination, as in the nickel complex of *N*-2-pyridylethylsalicylidenimine **VIII** (R = CH_2CH_2Py), which leads to the octahedral polyhedra $Ni_2N_4O_2$ [102]. The results described are very important to an understanding of the problem of the competitive coordination in a number of formally tridentate ligand systems: the various types of possible bondings depend upon the electronic configuration of the metals used, and upon the structures of the complexes formed by them [89,116–119].

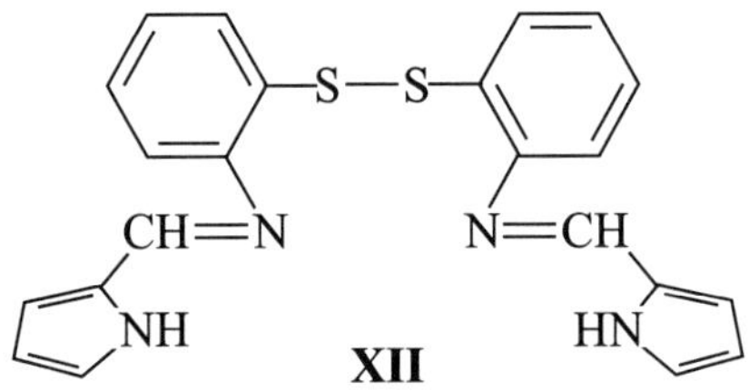

Azomethines containing an antipyrine fragment [107] on an N-atom could be attributed to the same ambidentate ligand systems. On the basis of this fragment, chelates of the type **XIII** (X = O), having the composition ML_2 where LH is the ligand, have been obtained [73] by the electrochemical method. In descriptions of these compounds both the participation and the nonparticipation of the O-atom from the carbonyl group in the coordination have been suggested [107,110], in addition to the participation of an oxygen atom from the phenolic group and a nitrogen atom from the azomethinic group. In this case, the electrochemical method has an important advantage: the complexes of composition ML_2 (**XIII**) are formed specifically, while in the usual synthesis conditions two types of coordination compounds (ML_2 (**XIII**) and MLAcO (**XIV**) [120]) have to be isolated by the use of metal acetates.

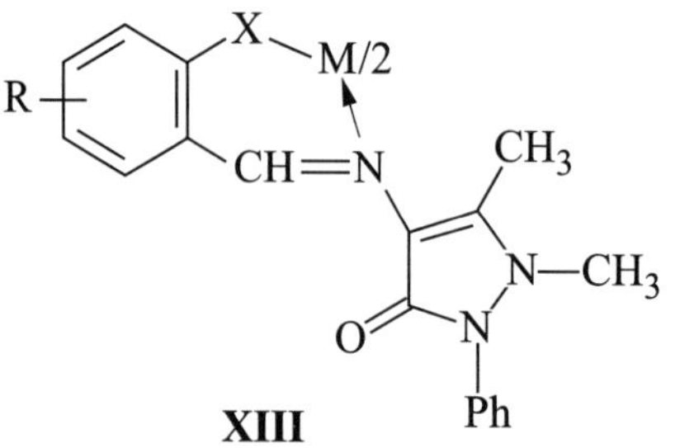

XIII 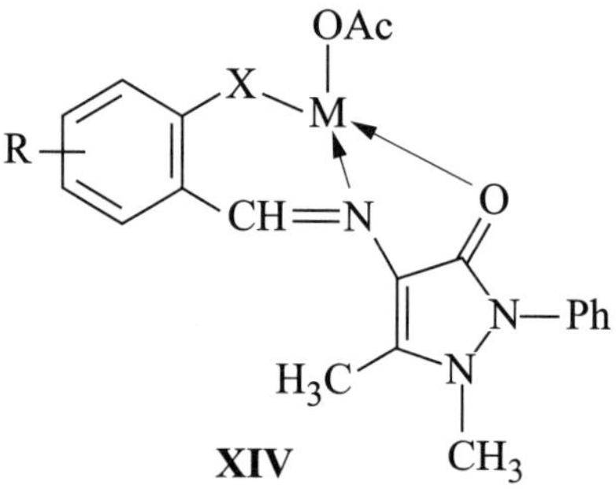   **XIV**

Formally six-dentate ligand systems **IX–XII** behave in an original way under conditions of electrosynthesis. For the ligand **IX**, the disulfide structure remains unchanged [105] during complex formation. The complex could be represented by the formula **XV** and includes a tetrahedral polyhedron, formed by the participation of the atoms of the phenolic oxygen and the azomethinic nitrogen atoms [105] in coordination with the metal.

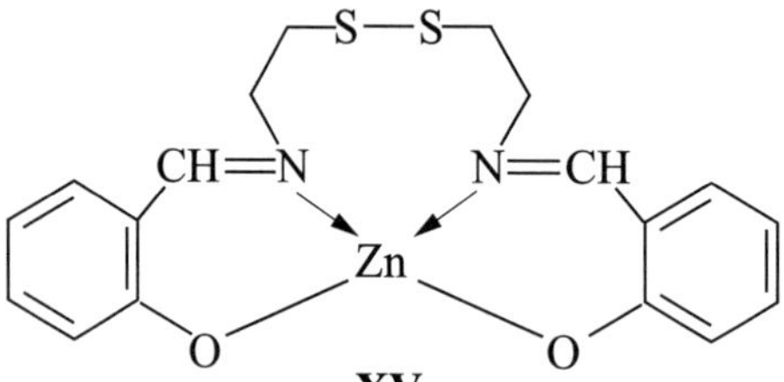

XV 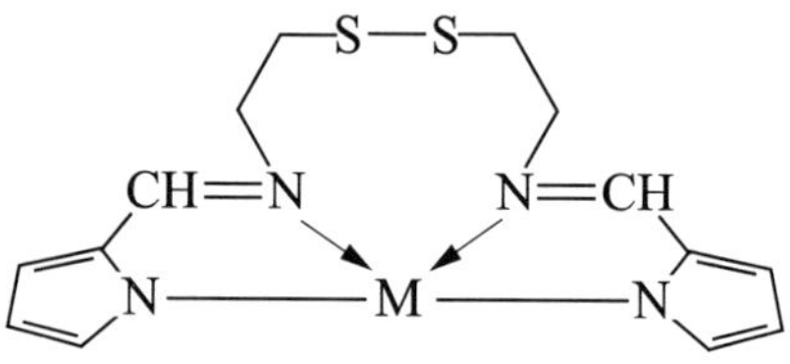   **XVI**

Complexes having the composition ML, based on the ligands **XI** and **XII** [104], for example the complex **XVI**, evidently have similar structures without the participation of S-atoms in the coordination. In the remaining cases, the cleavage of X–X bonds (X = S, Se) and the formation of complexes with the comparatively rare N,S- or N,O,S-ligand environment [89,116], are observed. It was confirmed by X-ray diffraction that two types of complexes, **XVII** (M = Cu, Ni, X = O [113]; M = Sn, X = S [106], Se [110]) and **XVIII** (M = Cu, Ni, L–L = o-phen; M = Zn, L–L = 2,2-bipy [108]; M = Co, L–L = o-phen [111]), are based on the basis of the compound **X**.

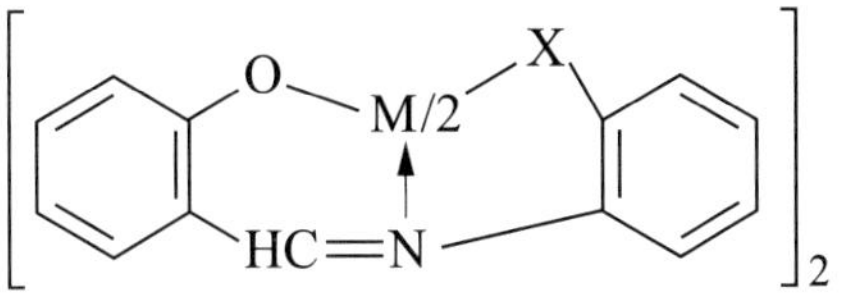

XVII

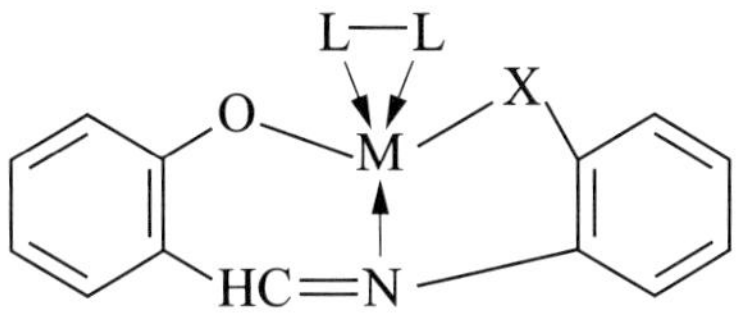

XVIII

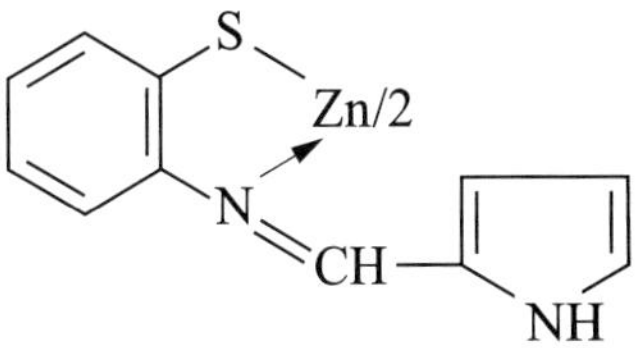

XIX

The complex **XIX** has the composition Zn(LH)$_2$; according to the X-ray diffraction data, it represents the complex of pyrrole-*o*-aminothiophenol, in which the pyrrole fragments do not take part in the coordination [104], and is formed by cleavage of the disulphide bond of ligand **XII** (LH$_2$).

The results described above extend the possibilities for the synthesis of complex compounds containing N- and chalcogen S(Se) donor atoms (recently a significant interest has been displayed in them [89,116–119]) in the coordination kernel. In this connection we note that complexes of the type **XIX**, containing five-membered coordination knots MN$_2$X$_2$ (X = S, Se) and aryl (hetaryl) substituents in the aldehydic fragment, were synthesized many years ago with the participation of metal salts or as a result of template synthesis [89,116].

The cycle of transformations examined above suggests that an electrochemical cleavage of an X–X bond (X = S,Se) [37,45,49,103,104,110] could open the way to synthesis of complexes with chelate five- and six-membered metal rings, containing a donor atom of tellurium (**XX** and **XXI**) (reaction (2.6)).

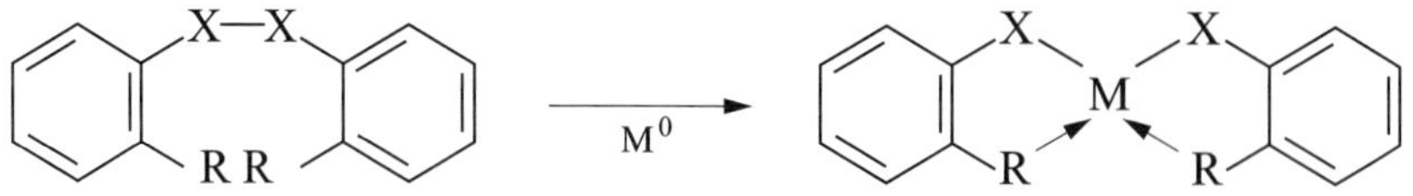

$$\textbf{XX} \qquad \qquad X = S, Se \qquad \qquad \textbf{XXI}$$

$$R = -NH_2, -NH\!=\!CH-Ar(Het), -CH\!=\!O, -CH\!=\!N-R^1 \text{ (Alk, Ar, Het)}$$

$$(2.6)$$

These studies, which take into consideration the elaborate methods of synthesis of the ligands **XX** [121,122] (R = $-NH_2$, $-N\!=\!CHR$, $-CH\!=\!O$, $-CH\!=\!N-Ar$), have been started by A. Sousa and co-workers (University of Santiago de Compostela, Spain) simultaneously with I.D. Sadekov and A.D. Garnovskii (State University of Rostov-on-Don, Russia). The possibility of obtaining complexes of the type **XXI** (X = Te) is confirmed by successful synthesis in the usual conditions (starting from ligands and metal salts, by template transformation) of chelates **XXI** (X = S, Se) [89,116].

The electrochemical synthesis of chelates with tridentate Schiff bases, based on 2-[(2-aminoethyl)-thiomethyl]benzimidazole, with a substituted acetophenone and acetylacetone, has been reported [115]. The study of the template electrosynthesis of the adducts of transition metal chelates using azomethinic ligands and benzimidazoles has been presented recently [123a]; a comparison of conventional chemical and direct electrochemical methods using azomethinic ligands as model compounds has been reported [124].

The coordination compounds **XXII–XXIV** with a five-membered metal ring MN_n have been obtained by an electrochemical method, based on azomethines derived from aldehydes with the five-membered nitrogen heterocycles pyrrole-2-aldehyde **XXII** [15,125–133], benzimidazole-2-aldehyde **XXIII** [17,21,98], as well as 2(2′-pyridyl)benzimidazole **XXIV** [10,17,21, 24,98]. Although the complexes **XXII** are the classic subjects of coordination chemistry [89,134, 135], the application of electrosynthesis not only makes it possible to extend the number of similar compounds, but also helps to specify their structures, and the problems of competitive reactions of azole ligand systems with zero-valent metals.

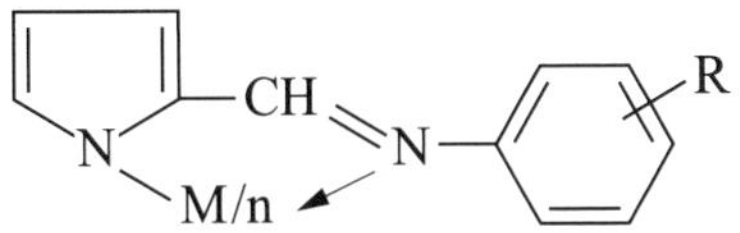

XXII

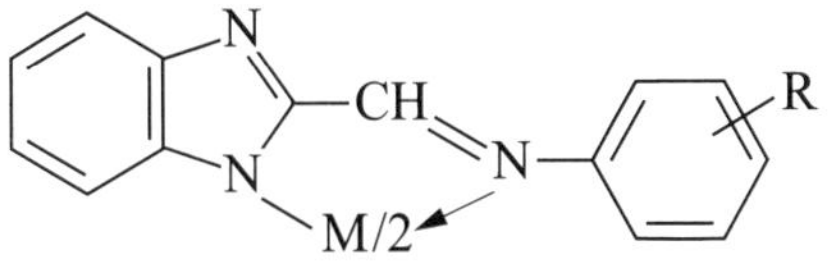

XXIII

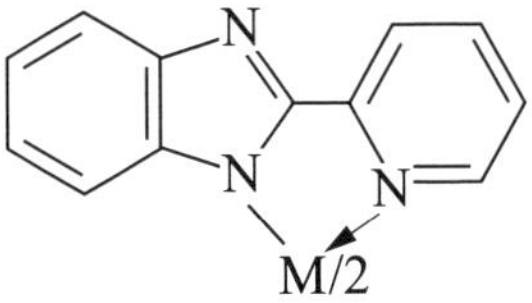

XXIV

The X-ray diffraction data for pyrrolealdiminates **XXII**, obtained under electrosynthesis conditions (AN solution), testify that the copper complex **XXII** (M = Cu^{2+}, R = p-Me, $n = 2$) [126] includes a tetrahedally distorted metal-containing ring CuN$_4$, but the chelate of three-valent cobalt **XXII** (M = Co^{3+}, R = o-Me, $n = 3$) [127] represents an octahedron with the coordination knot CoN$_6$. The analogous ligand enviroment N$_6$, as the results of X-ray diffraction testify, is characteristic of the adducts of nickel chelates **XXII** (M = Ni^{2+}, R = H, $n = 2$) and cadmium chelates **XXII** (M = Cd^{2+}, R = 2-OMe, n = 2) [128] with α, α'-bipy. For the ambidentate ligand systems of the pyrrolealdimine type, the competitive coordination is very characteristic, with the formation of three types of structures: with participation of a phenolimine fragment as in **XXV** [15,129,133], of all the donor atoms as in **XXVI** [130], and of imino-amine groups as in **XXVII** [132].

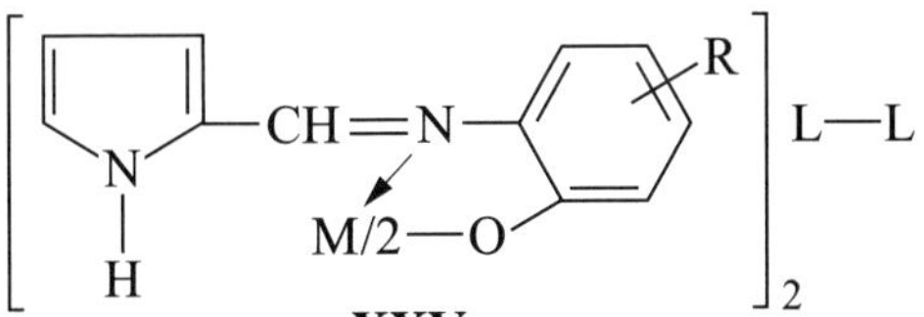

XXV

R = H, 4-Me, 5-Me, 5-Cl, 3,5-dimethyl
M = Cd, Ni, Zn; L–L = bipy, o–phen

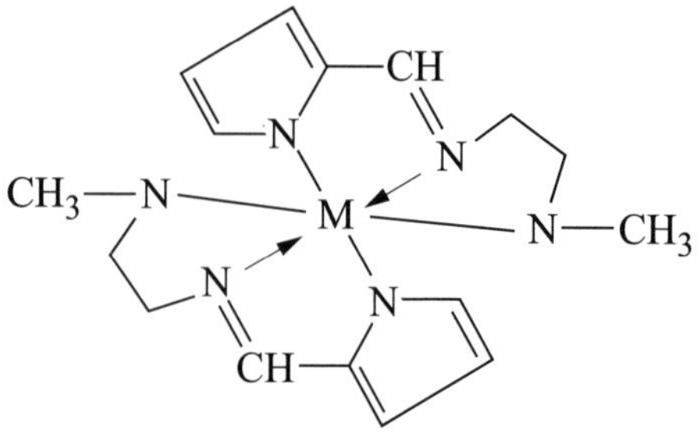

XXVI

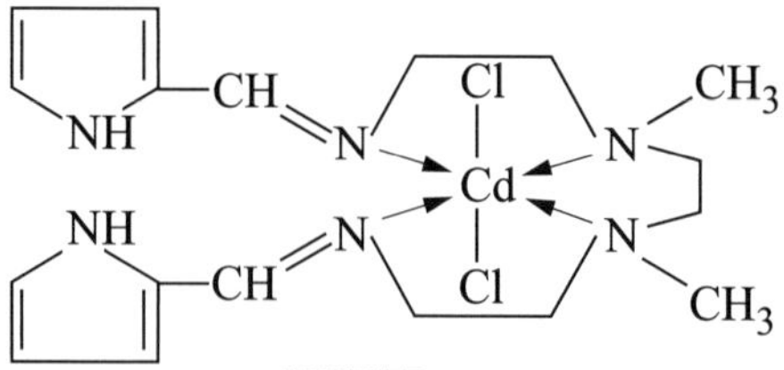

XXVII

We emphasize that the pyrrole fragment does not take part in coordina-tion in the complexes **XXV** and **XXVII**. This situation may be connected with the conduction of the electrosynthesis procedure in acetonitrile solution in the presence of such strong chelating N-donors as 2,2′-bipy, 1,10-phen (the complexes **XXV**) or macrocyclic fragments (the complex **XXVII**). Confirmation is the fact that electrochemical synthesis based on ligands of the azole series with less NH-acidity than pyrrole [136], in alcohol solution, also leads to the *N*-metal-substituted derivatives **XXIII** and **XXIV** [10,17,21,24,98]. It is very characteristic that *N*-metal substitution is also observed during electrochemical synthesis in methanol or ethanol for azoles themselves [10,17,21,24,98].

Among other chelates with a five-membered metal ring, we also note the complexes synthesized by an electrochemical method based on bidentate oxygen-containing donors such as **VI** (R = Me) [137a], **XXVIII** [137b] and pyridine-2-thiol-*N*-oxide **XXIX** [138a].

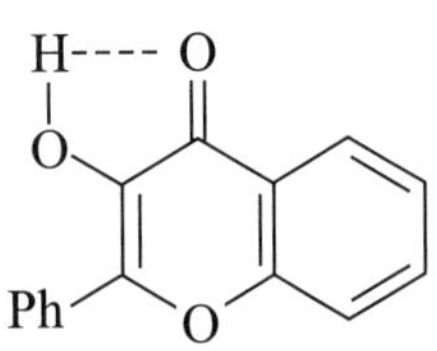

XXVIII 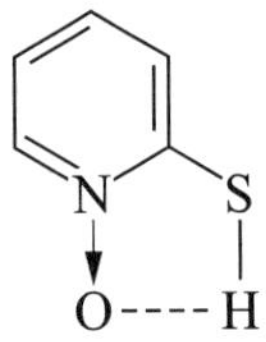   **XXIX** 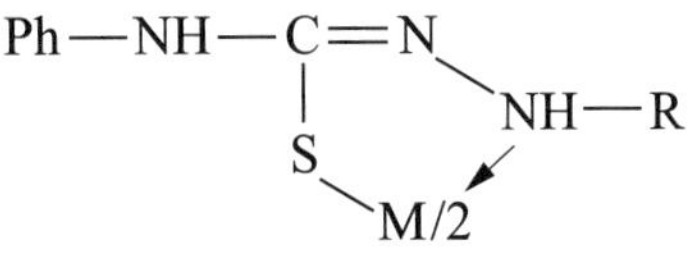   **XXX**

The existence of five-membered metal rings of the types MO_4, MO_2S_2 and MS_2N_2 in chelates of the ligands **VI** and **XXVIII**, **XXIX** and, **XXX**, respectively, is confirmed by X-ray diffraction [137,138].

Five-membered metal rings are found in the chelates of hetaryl-substituted monosugars such as **XXXI**, having the composition ML_2, which were prepared with a yield of 80–90% by electrolysis in methanol solution of the ligands with an anode made from the corresponding metal (cobalt, nickel, copper) and a platinum cathode (see [2], p. 1026). This is an advantage of the electrochemical route, since in the usual conditions of synthesis starting from LH and $M(Acet)_2$ (Acet = acetate anion), the complexes have the compositions ML_2 and MLAcet and are formed in only 30% yield [2].

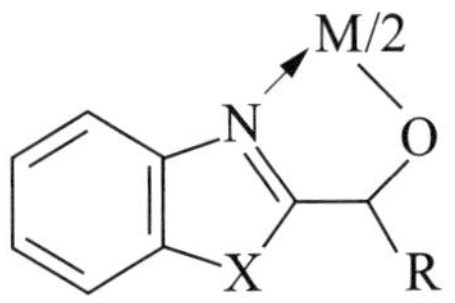

XXXI   **XXXII**

$X = NCH_3$, O; R = glycoside

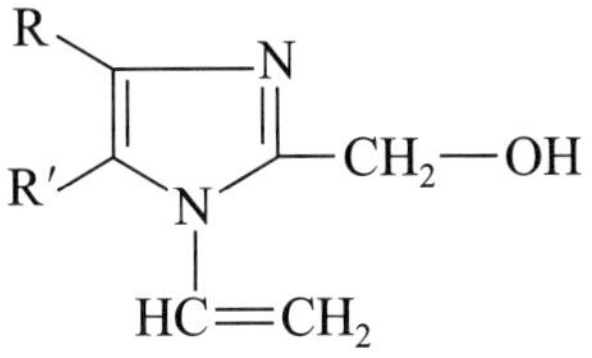

Highly unexpected results were obtained from electrochemical synthesis for coordination compounds of 1-vinyl-2-hydroxymethyl(benz)imidazoles (LH) **XXXII** [139]. For those reactions conducted in acetonitrile solutions, with electrodes made from 3d metals (Ti,Cr,Fe,Cu) in the presence of Bu_4NClO_4, instead of the expected complexes with composition ML_n, the chelates retaining the perchlorate anion $LM(ClO_4)_n$ were produced, or the mixed-ligand chelates $(LH)LM(ClO_4)_n$ in the case of titanium and copper, as well as the binuclear compounds $LCu_2(ClO_4)_3$ [139].

Chelates with four-membered chelate knots obtained electrochemically are represented by the complexes of o-diphenols **XXXIII** [67,68,140,141], alkanedithiols **XXXIV** [76,77] and their adducts with N and O bases (chelates with a large number of sections with metal-containing rings are also described in the same publications). Thus, complexes of composition ML (LH$_2$ = ligand) **XXXIV** [76] or InLH, InI$_2$·LH (with mono-H-substituted ligand; in the last case in the presence of iodine) **XXXV**, as well as the anionic complexes [NR$_4$][InL] and [NR$_4$][InI$_2$L], have been synthesized [77] (see Table 2.2).

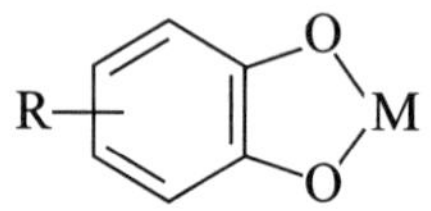

| **XXXIII** | **XXXIV** | **XXXV** |

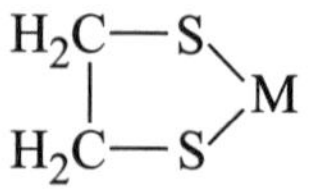

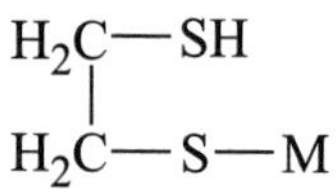

Electrochemical synthesis has been used to obtain the carboxylates [80], diethyldithiocarbamates **XXXVI** [142], diethyldithiophosphates **XXXVII** [142], chelates of pyridine-2-ol **XXXVIII** (X = S, A = CH) [143], pyrimidine-2-thiol **XXXVIII** (X = S, A = N [144,145]) and their molecular adducts with azines and bisazines [144b,145].

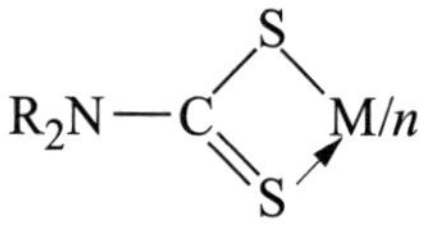

| **XXXVI** | **XXXVII** |

| **XXXVIII** | **XXXIX** |

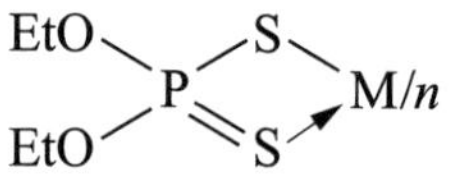

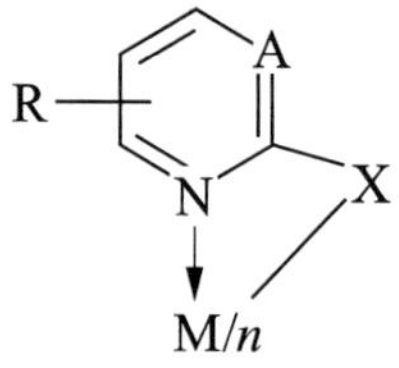

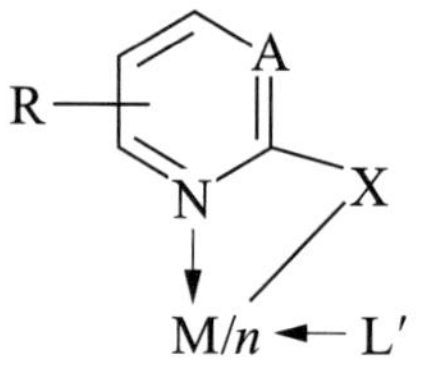

The electrosynthesis of dialkyldithiocarbamates **XXXVI** (R = Me, Et) and diethyldithiophosphates **XXXVII** was carried out [142] in acetonitrile solution with anodes made of copper, silver, zinc, cadmium, indium, thallium, cobalt, nickel and iron. The adducts ML$_n$·phen (M = Ga, V, Mn, Co, Fe) were obtained by addition of 1,10-phen to the reaction mixture. The complexes were characterized by means of analytical data, which supported [142] the formation of chelates with the compositions ML_2 (**XXXVI, XXXVII**; $n = 2$) and ML (**XXXVI** (M = Ag, Tl; $n = 1$), **XXXVII** (M = Cu,Tl,Ga; $n = 1$), as well as ML_3 (**XXXVI** (M = Fe, In, Tl; $n = 3$), **XXXVII** (M = Co; $n = 3$). The ^{31}P NMR spectra of **XXXVII** have been discussed [142]. In the case of the copper–tetramethylthiuram disulfide (TMU) system, a detailed study [146] of the interaction of metallic copper with nonaqueous solutions of TMU in the presence of CI_4 and bipy without use of electrosynthesis must be mentioned. The compounds formed have the general formula $[Cu_nI_m(Me_2NSC_2)_l]\cdot A$, where $n = 1–3$, $m = 0$, 1, 2, 4, $l = 1–3$, 5; A = H_2O, Cu_2S and CHI_3, containing copper in the forms Cu(I), (II) or the rare Cu(III), according to the magnetic data.

The metal chelates of two-valent iron, copper, zinc and cadmium [147], and cobalt and nickel [148] based on pyrid-2-one **XXXVIII** (A = CH, X = O) were isolated in electrosynthesis conditions. The results of magnetic measurements, in our opinion, are of the greatest interest among the physicochemical properties of complexes **XXXVIII** and **XXXIX** (X = O, A = CH). In particular, the magnitudes of μ_{eff} for the copper complexes $CuL_2\cdot3H_2O$ and $[CuL_2\cdot phen]\cdot3H_2O$, are 1.5 and 2.4 μ_B, respectively [147a], and could be linked with an exchange interaction between the antiferromagnetic and ferromagnetic types.

It is well known that the ligands of hetaryl-2-ones and hetaryl-2-thiones form several types of complexes [70,143a,144a,145b]. The X-ray diffraction data testify that for the adducts of pyridine-2-thiol **XXXIX** (A = CH, X = S, M = Ni [143a], $L' = $ bipy, R = H, $n = 2$; M = Zn [143b]; $L' = $ phen, R = H, $n = 2$) and pyrimidine-2-thiol **XXXIX** (A = N, X = S; M = Ni [144a], $L' = $ bipy, $n = 2$; M = Cd [144b], $L' = $ phen, R = H, $n = 2$; M = Zn [145a], $L' = $ Py, R = 4,6-dimethyl, $n = 2$; M = Ni, Cd [145b], $L' = $ bipy, phen, $n = 1$), the structures are very charactersitic, with the four-membered metal ring MN_2S_2. This ring is formed by an endocyclic nitrogen atom of the pyridine type and the sulfur atom of a thiol. The adduct-forming ligands (L') form the octahedral (bipy, phen) [143–144, 145b] or trigonal-bipyramidal (Py) [145a] configurations.

There is at least one good reason to think that the electrosynthesis could be used to obtain macrocyclic chelates, in particular of the porphyrin or phthalocyanine series [24]. The feasibility of such a process has been reported by Yang and Straughan, who obtained some metal phthalocyanines electrochemically from metal salts (Cu, Pb, Mg, Co, Ni) and elemental Fe and Cu

[149]. Furthermore, Petit's research group [150] studied the electrosynthesis of copper phthalocyanine from the electroreduction of phthalonitrile using as the anode a copper sheet or an electrodeposited layer of copper on platinum. They reported effect of some of the electrolytic parameters on the yields of the process and a mechanism involving several steps in this paper. Another important result from this group is the electrosynthesis of the lithium phthalocyanine radical which is known to be a member of the class of intrinsic molecular semiconductors [151,152]. In our opinion, the electrosynthesis of phthalocyanines and related macrocycles is a very promising field and it should be developed in future investigations (see the data on the metal dissolution in systems containing the precursors of phthalocyanine in Chapter 3).

Thus, the main attention of the research groups working in the electrosynthetic area has been devoted to its most productive subarea: isolation and characterization of various transition metal chelates. Summarized data on the electrosynthesis of metal chelates are presented in Table 2.5.

Experimental Procedure

The experimental procedures used in the electrosynthesis of some of the complexes reported in Table 2.5 are described in the following examples.

(a) Electrosynthesis of transition metal complexes with heteroazomethinic ligands [124]

The electrosynthetic processes were carried out by anodic dissolution of metals in acetonitrile or methanol solutions of the corresponding ligands, according to the literature [14a,14b,18,123].

The electrolytic cell consisted of a tall beaker, and the solution phase contained the appropriate ligand (1–2 g) dissolved in acetonitrile, or in methanol containing tetrabutylammonium bromide or sodium chloride ($0.1 \, mol \, cm^{-3}$). The anode (sacrificial metal) was supported on a platinum wire and a platinum sheet formed the cathode. All experiments were conducted at room temperature in an atmosphere of dry nitrogen, which was bubbled through the cell. The voltage was regulated to obtain an initial current of 20 mA. In each case the electrolysis was conducted for 2–2.5 h, after which the anode was washed, dried and weighed in order to obtain the electrochemical efficiency. The experimental conditions are presented in Table 2.6. The yields of the final products were 85–96%.

In the case of copper complexes some of the product was occasionally formed on the surface of the anode, which led to an increase in the voltage. To avoid this, a combination of electrolysis with simultaneous treatment by ultrasound was applied.

TABLE 2.5
Electrosynthesis of metal chelates

Metal	Initial system	Products	Ref.
Co, Dy, U	EtOH, Hba or HDBM	Co(EtOH)(ba)$_2$] Co(DBM)$_2$·3EtOH [Dy(HDBM)(DBM)$_3$]	95–97
	In N$_2$ atmosphere	U(ba)$_4$	
	In the presence O$_2$	UO$_2$(ba)$_2$(Hba)$_{0.5}$	
In, Fe, Ni, Ti, Zn, Zr, Cu, Mn, Co V, Hg, Th, Sm, Gd, U	Hacac, CH$_3$OH, N$_2$ Hacac, O$_2$	M(acac)$_n$ UO$_2$(acac)$_2$ UO$_2$(acac)$_2$·Hacac VO(acac)$_2$	27, 64, 66, 72, 91, 93, 95
Th	Acetone, tfha	Th(tfha)$_4$	27
Cd, Co, Ni, Zn	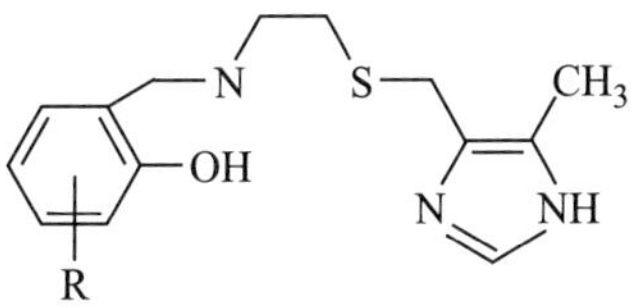 [structure: 2-hydroxybenzyl-N–CH$_2$CH$_2$–S–CH$_2$– linked to 4,5-dimethylimidazole; ring bearing OH and R] R = H, 3-OEt, 5-Br, CH$_3$CN	ML·nH$_2$O	153
Cd, Co, Ni, Zn	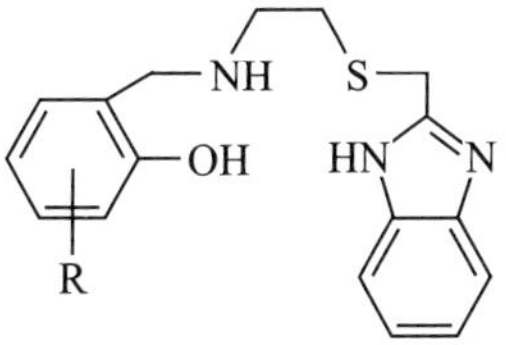 [structure: 2-hydroxybenzyl-NH–CH$_2$CH$_2$–S–CH$_2$– linked to benzimidazole; ring bearing OH and R] R = H, 3-OEt, 5-OMe, CH$_3$CN	ML·nH$_2$O	153
Cd, Cu, Ni, Zn	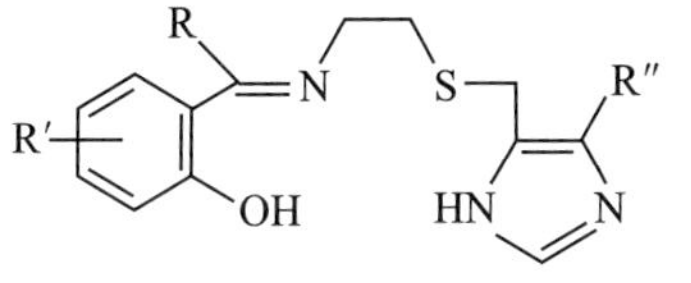 [structure: salicylidene R-C=N–CH$_2$CH$_2$–S–CH$_2$– linked to imidazole bearing R″; ring bearing OH and R′] R = H; Me R′ = H; 5-OMe; 4,6-(OMe)$_2$; 5-Br; 3,5-Br$_2$; R″ = H, Me, CH$_3$CN	ML·nH$_2$O	154

TABLE 2.5   (contd.)

Metal	Initial system	Products	Ref.
Cd, Ni, Zn	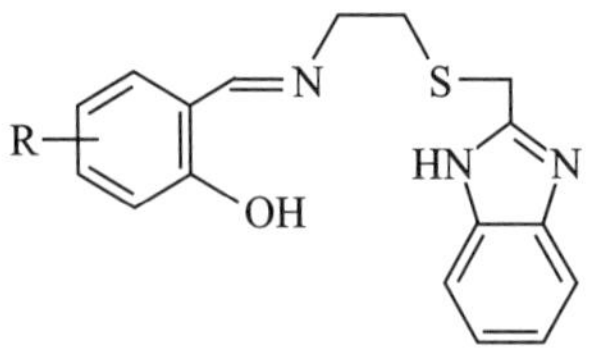 R = H; 3-OEt, 5-OMe; 4,6-(OMe)$_2$; 5-Br CH$_3$CN	ML·nH$_2$O	154
Cu, Ni	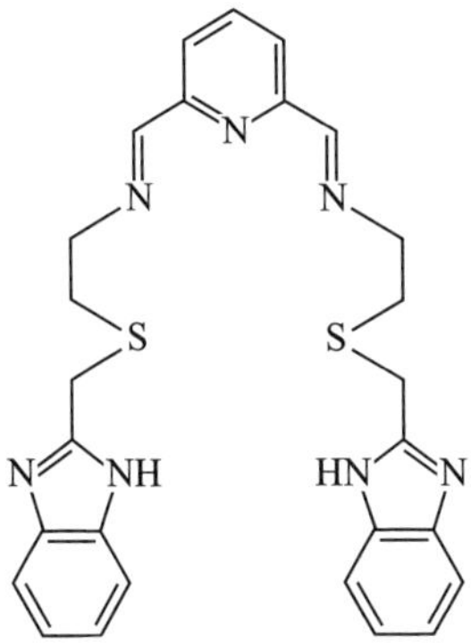 (bH-pyrathH$_2$), EtOH	[M(bHpyratb)]·nH$_2$O	115
Cd, In, Hg, Sn	 CH$_3$CN, toluene In the presence of bipy	M(C$_{15}$H$_9$O$_3$)$_n$ Cd(C$_{15}$H$_9$O$_3$)$_2$·bipy	137a
In, Zn, Cd	Toluene, CH$_3$CN, 3-hydroxy-2- methyl-4-pyrone (= HL)	ML$_n$	137b
Cd	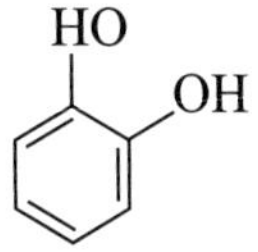 CH$_3$CN, toluene	Cd(C$_8$H$_9$O$_2$)$_2$	137a

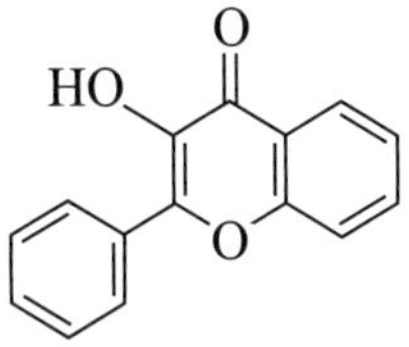

TABLE 2.5 (contd.)

Metal	Initial system	Products	Ref.
Cd, Zn, Cu	Acetone, HPTS	$[Cu(PTS) \cdot H_2O](HPTS)$ $[Cd(PTS)_2 \cdot 2H_2O] \cdot H_2O$ $Zn(PTS)_2(HPTS)$	138b
	In the presence of phen	$[Cd(PTS)_2 \cdot phen] \cdot 2H_2O$	
Zn	$H_2L = $ 1,7-bis(2-benzimidazolyl)-2,6-dithiaheptane, CH_3CN or 1,6-bis(2-benzimidazolyl)-2,5-dithiahexane, EtOH	$[Zn(bbdhp)] \cdot 3.5H_2O$ $[Zn(bbdhx)] \cdot 3.25H_2O$	155
Cd	2,2-Dipyridylamine (= HL), CH_3CN	CdL_2	36
Cd, Zn	C_6H_6, CH_3OH, HL (L = oxine, 8-hydroxyquinoline), Hsal (salicylaldehyde)	ML_2	91
Co	Nonaqueous solvent, SAAP and NAAP (= ROH)	$\underline{ROH = SAAP}$ $Co(OR)_2(H_2O)_2$	107
	In the presence of phen	$Co(OR)_2 \cdot phen \cdot 3H_2O$	
	In the presence of Ph_3P	$Co(OR)_2 \cdot Ph_3P \cdot H_2O \cdot 5H_2O$ $\underline{ROH = NAAP}$ $Co(OR)_2 \cdot 2H_2O$	
	In the presence of phen	$Co(OR)_2 \cdot phen \cdot 2H_2O$	
	In the presence of Ph_3P	$Co(OR)_2 \cdot Ph_3P \cdot H_2O \cdot 4H_2O$	
Co, Cu, Ni	CH_3OH, azoles or their derivatives (HL)	CoL	10–13
Cu, Ni	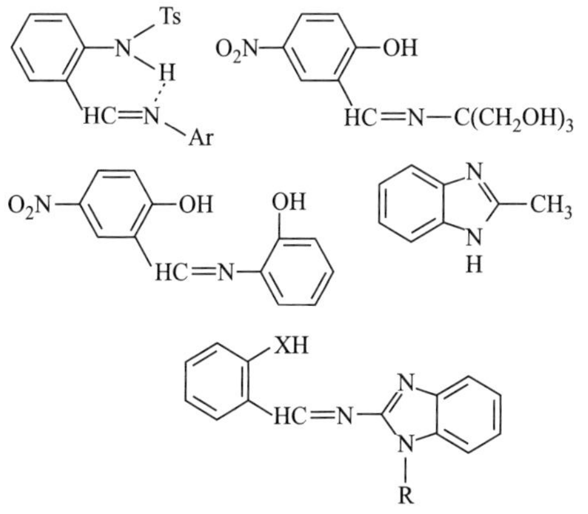	CuL_n or $CuL_n \cdot Solv$	124, 156–158

CH₃CN or EtOH

TABLE 2.5   (contd.)

Metal	Initial system	Products	Ref.
Ni, Cu, Co, Zn, Cd	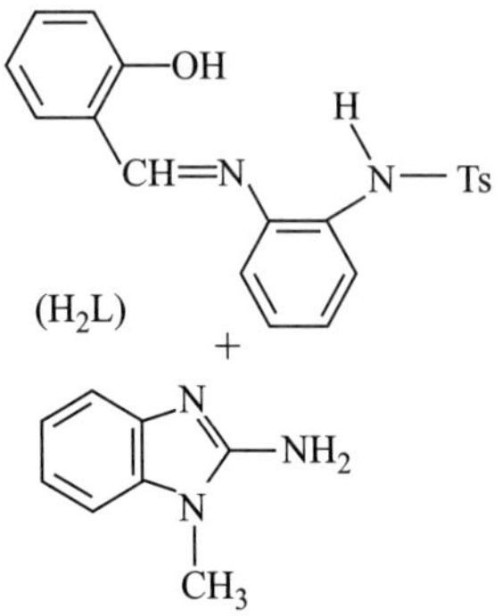 (H$_2$L) +	ML·mL′·nH$_2$O $m = 1$ (M = Ni, Cu) $m = 2$ (M = Co, Zn, Cd) $n = 0$–2 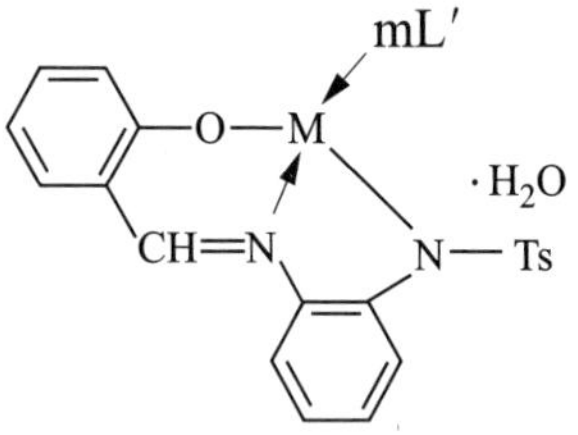	123a
Ni, Cu, Zn, Cd	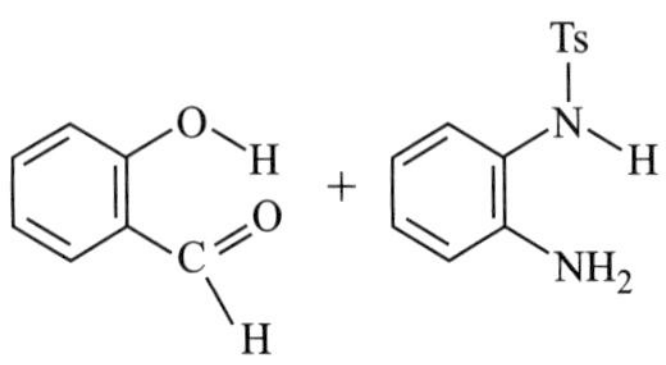	See example (e) in Experimental procedure, Section 2.3.2	123b
Zn, Cd	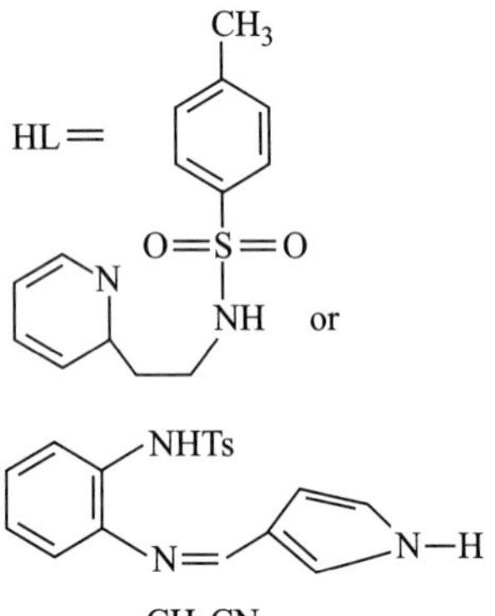 HL = or CH$_3$CN	ML$_2$	159, 160
Cu	Schiff bases, derived from 2-(2-aminoethyl)pyridine and substituted salicylaldehydes (HL), CH$_3$CN	CuL$_2$	100

TABLE 2.5 (contd.)

Metal	Initial system	Products	Ref.
Cu	EtOH, CH_3ONa, phthalonitrile	CuPc	150
Fe, Cu	CH_3OH, 1,2-dichlorobenzene	CuPc, FePc	149c
La, Sm, Nd, Pr	*i*-BuOH, CH_3ONa, phthalonitrile	M_2Pc_3	161
Cd, Co, Cu, Fe, Ni,	CH_3CN or acetone, $(S_2CNR_2)_2$ (R = Me, Et)	$M(S_2CNR_2)_2$	142
Zn	CH_3CN, $(S_2CNEt_2)_2$	$Tl(S_2CNR_2)_3 + Tl(S_2CNR_2)$	
Tl		$Au(S_2CNEt_2)$	
Au		$In(S_2CNR_2)_3$	
In			
Cd	R_2NH (R = Me, Et, *i*-Pr, piperidine); CS_2	$Cd(S_2CNR_2)_2$	142

Hba, benzoylacetone; HDBM, dibenzoylmethane; tfha, 1,1,1-trifluoromethylheptane-2,5-dione; HPTS, 4-phylthiosemicarbazide; bbdhp, 1,7-bis(2-benzimazolyl)-2,6-dithiaheptane; bbdhx, 1,6-bis(2-benzimidazolyl)-2,5-dithiahexane; Hsal, salicylaldehyde; SAAP, 4-aminoantipyrine Schiff base of Hsal; NAAP, 2-hydroxy-1-naphthaldehyde; Pc, phthalocyanine

This stabilized the voltage because of a faster elimination of the complex formed on the anode surface [156].

(i), (ii)

Ts = —SO_2—C_6H_4—CH_3

(iii), (iv)

(v)

(vi), (vii)

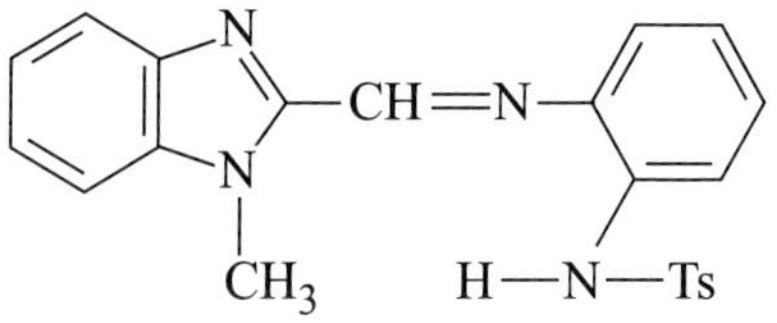

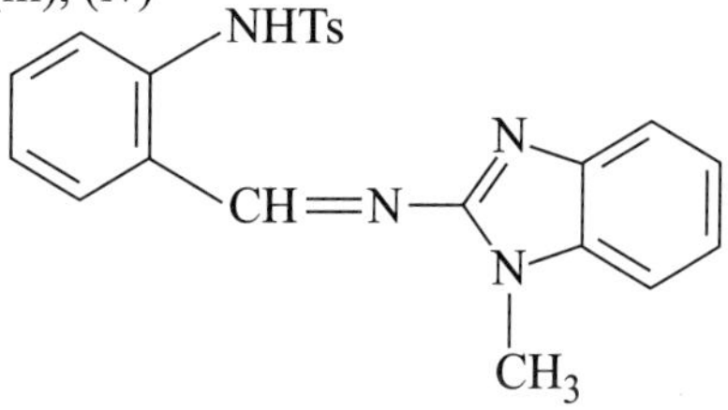

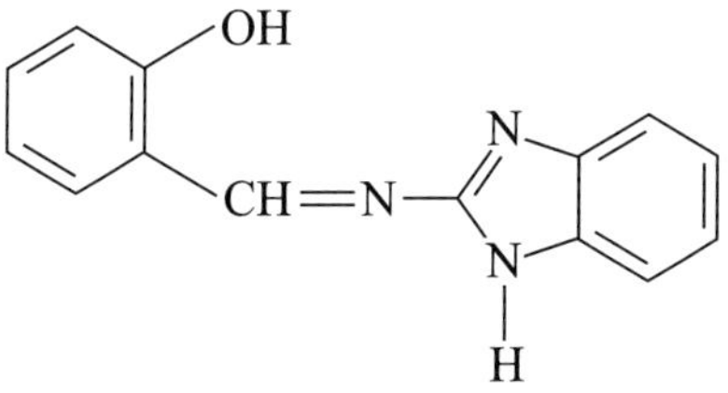

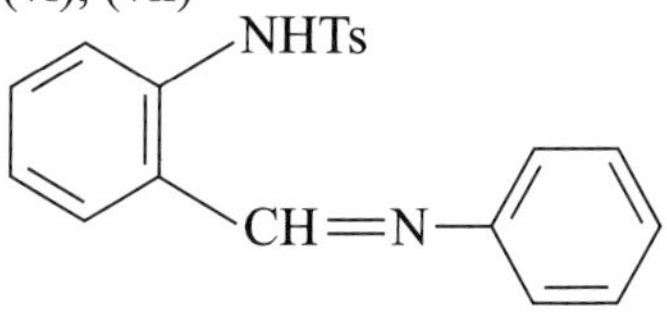

TABLE 2.6
Experimental conditions for the electrosynthesis of metal chelates

Compound		Metal	Initial voltage V*	Time of electrolysis, h	Metal dissolved, g	Yield, g/%
(i)	CuL·2H$_2$O	Cu	55	2	0.092	0.64/95
(ii)	NiL$_2$·2.5H$_2$O	Ni	62	3	0.048	0.72/96
(iii)	CuL·2H$_2$O	Cu	85	1.2	0.051	0.40/96
(iv)	NiL$_2$	Ni	71	3	0.065	0.82/85
(v)	CuL·H$_2$O	Cu	48	2	0.093	0.44/94
(vi)	CuL$_2$	Cu	58	2	0.045	0.52/96
(vii)	NiL$_2$	Ni	46	3	0.062	0.75/94
(viii)	CuL$_2$·3H$_2$O	Cu	65	2	0.047	0.50/92

*To produce a current of 20 mA.
The azomethinic ligand formulae (complex precursors) used are the following:

(viii)

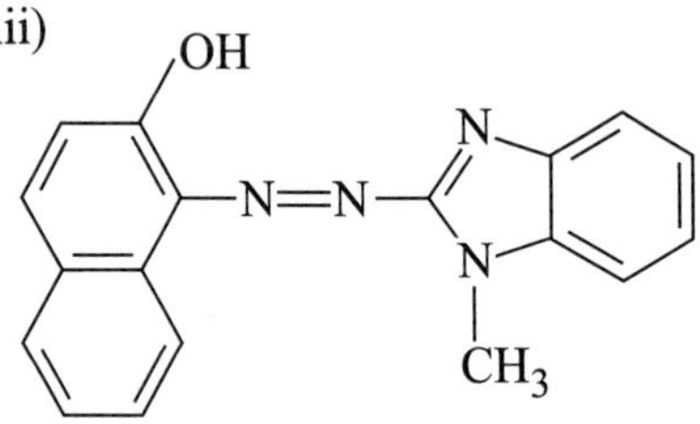

(b) Electrosynthesis of bis{4-methyl-N-(2-pyridin-2-ylethyl)benzenesulfonamide}copper(II) and bis{4-methyl-N-(2-pyridin-2-ylethyl)benzenesulfonamide}nickel(II) [159]

The ligand HL was prepared by the reaction of equimolar amounts of 2-aminoethylpyridine and tosyl chloride in dichloromethane and its purity was checked by conventional spectral analysis.

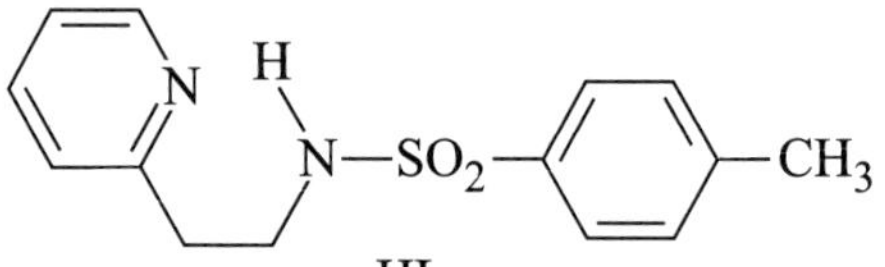

HL

The electrochemical technique was similar to those described by Tuck and co-workers [91]. The cell was a 100-cm^3 tall beaker fitted with a rubber bung through which the electrochemical lead entered the cell. A platinum wire suspended the sacrificial metal and served as anode. The tosyl derivative was dissolved in acetonitrile, and a small amount (ca 20 mg) of tetraethylammonium perchlorate was used as supporting electrolyte.

The anodic oxidation of Cu metal in a solution of the ligand (0.2 g) in acetonitrile (50 cm^3) for 2 h at 20 mA resulted in 94 mg of anode loss; $E_F = 10$ mol F^{-1}. The reaction mixture was filtered to remove any precipitated particles of metal, and the filtrate was left to concentrate by evaporation at room temperature.

A crystalline product identified as [CuL$_2$] was isolated. Found (%): C, 54.5; H, 4.6; N, 9.1; S, 9.3. Calculated (%) for C$_{28}$H$_{30}$N$_4$O$_4$S$_2$Cu: C, 54.7; H, 4.9; N, 9.1; S, 10.4.

The electrochemical oxidation of Ni in a solution of the ligand (0.42 g) in acetonitrile (50 cm^3) for 2 h at 20 mA led to 45 mg of anode loss; $E_F = 0.51$ mol F^{-1}). The compound identified as [NiL$_2$] was isolated by evaporation of the solution. Found (%): C, 55.0; H, 5.2; N, 9.3; S, 10.7. Calculated (%) for C$_{28}$H$_{30}$N$_4$O$_4$S$_2$Ni: C, 55.2; H, 5.0; N, 9.2; S, 10.5.

(c) Electrosynthesis of phthalocyanines [150]

The electrolysis cell (see Fig. 2.3) was a divided jacketed three-electrode cell. The cathode was a gold grid (2 cm × 2 cm, 196 mesh per cm^2) whereas the anode was a twisted platinum wire.

A 100-cm^3 solution of the chosen solvent containing LiCl (3 g, 0.07 mol) was introduced into the cell and deaerated at a given temperature. The electrolytic process was started after the addition of PN to the cathodic compartment. With EtOH as the solvent, the initially colorless solution became yellow, then blue green after the passage of 20–40 C and it finally turned into a viscous blue suspension. The electrolysis was stopped after a given amount of charge had been passed. Then, the catholyte was poured into 100 cm^3 of a 0.2 mol dm^{-3} H$_2$SO$_4$ solution. The resulting suspension was stirred for 0.5 h and then filtered. The blue solid was treated suitably and the yields were calculated.

Elemental and spectroscopic analyses indicated that the blue solid was the hydrogen phthalocyanine PcH$_2$ (α-form).

The same electrolytic process was applied to a substituted phthalonitrile.

The electrosynthesis of PcCu was performed using a copper sheet as anode or an electrodeposited layer of copper on platinum. CuClO$_4$ was introduced into the anodic compartment.

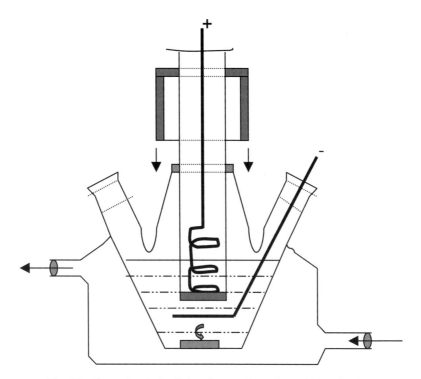

Fig. 2.3. Electrochemical cell (a reference electrode could be added) [150]

(d) Electrosynthesis of zinc and cadmium complexes with N-[(2-pyrrolyl)methylidyne]-N'-tosylbenzene-1,2-diamine [160]

Acetonitrile and pyrrole-2-aldehyde were used without further purification. N-Tosyl-1,2-diaminobenzene was obtained following the method described by Malick and co-workers. The Schiff base N-[(2-pyrrolyl)methylidyne]-N'-tosylbenzene-1,2-diamine (H_2L) was isolated from the condensation of N-tosyl-1,2-diaminobenzene and pyrrole-2-aldehyde in methanol.

The electrolysis used in the synthesis of complexes was similar to that described by Tuck [51]. The cell was a tall beaker with a rubber bung through which the electrochemical leads entered the cell. The metallic anode was suspended from a platinum wire in a solution of the corresponding ligand in acetonitrile. Another platinum wire was used as cathode.

$[Zn(HL)_2]$ was obtained by electrolysis of an acetonitrile solution ($50\,cm^3$) containing H_2L (0.25 g, 7.35 mmol) and a small amount of tetramethylammonium perchlorate (ca 10 mg), at 20 V and 10 mA for 4 h, by which time 48 mg of zinc had dissolved ($E_F = 0.98 \text{ mol F}^{-1}$). After the experiment the yellow solid was filtered, washed with hot acetonitrile and ether, and dried in vacuo. The compound was characterized as [Zn(HL)$_2$] by elemental analysis. Found (%): C, 58.1; H, 4.4; N, 11.4; S, 8.4. Calculated (%) for [C$_{36}$H$_{32}$N$_6$O$_4$S$_2$Zn]: C, 58.5; H, 4.1; N, 11.4; S, 8.7.

[Cd(HL$_2$) was prepared by electrochemical oxidation of cadmium in an acetonitrile solution (50 cm^3) containing the ligand H$_2$L (0.25 g, 7.35 mmol) using a 10 mA current for 4 h, which resulted in an anode loss of 92 mg ($E_F = 1.08 \text{ mol F}^{-1}$). After the experiment the yellow solid was treated as described above. Found (%): C, 54.4; H, 4.1; N, 10.7; S, 8.0. Calculated (%) for [C$_{36}$H$_{32}$N$_6$O$_4$S$_2$Cd]: C, 55.0; H, 4.1; N, 10.7; S, 8.1.

(e) Ni, Cu, Zn, Cd + salicylaldehyde + 2-N-tosylaminoaniline (first example of template electrosynthesis of metal chelates) [123b]

Metal sheets (2 cm × 2 cm) and platinum wire were used as anode and cathode, respectively. The electrolysis was carried out in CH$_3$CN at 10 mA for 4 h. Et$_4$NClO$_4$ was used as supporting electrolyte. Salicylaldehyde (0.122 g, 10^{-3} mol) and 2-N-tosylaminoaniline (0.262 g, 10^{-3} mol) were used as ligand precursors. The ligand is formed from its precursors by the reaction:

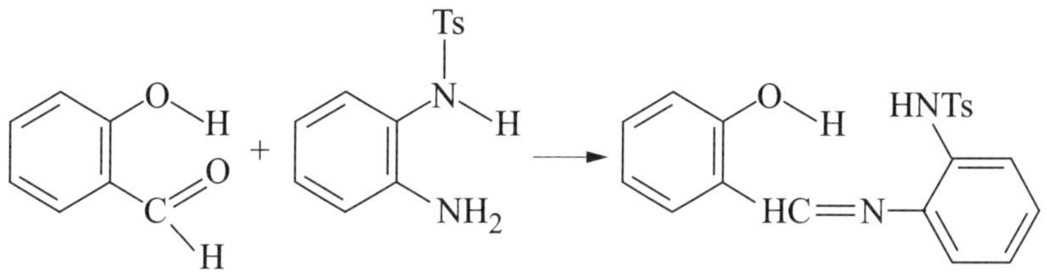

Complexes of general formula ML·mL′·nH$_2$O, where $m = 1,2$; $n = 0$–2, were precipitated, filtered off, washed with hot acetonitrile and methanol, and dried in vacuo at 70°C.

It is shown that the character of the metal complexes obtained depends on the nature of the metal.

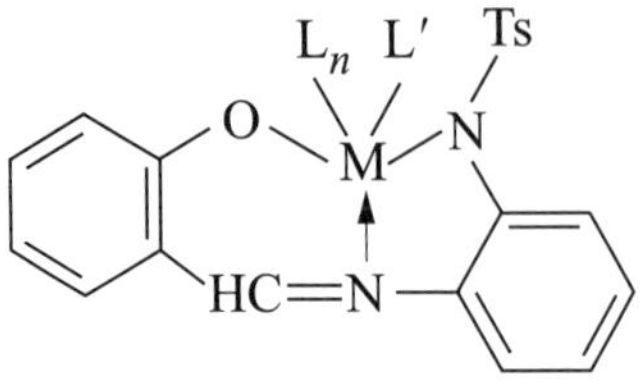

M = Cu, L = H$_2$O, n = 1

M = Ni, Cd, Zn; n = 0

L′ = 1-methyl-2-aminobenzimidazole

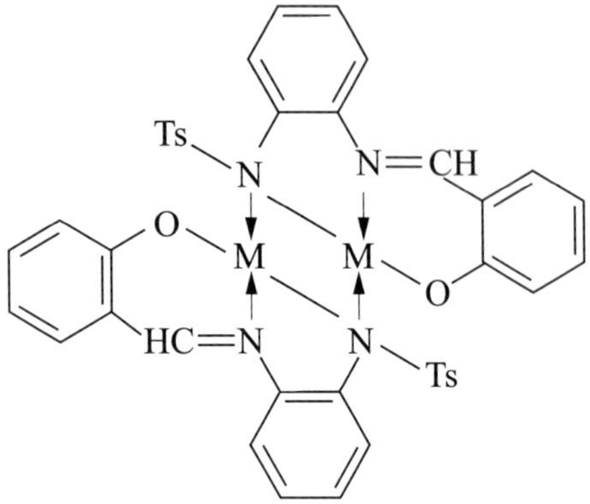

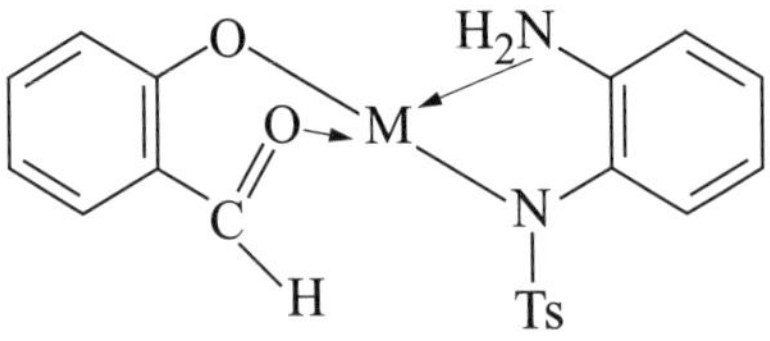

M = Ni

(f) Synthesis of iron(II) acetylacetonate [64]

NaCl (2 g) was dissolved in a mixture of 60 cm^3 of water and 50 cm^3 of methanol with 40 cm^3 of acetylacetone. This electrolyte was electrolyzed at 25°C between two iron electrodes. The electrolytic process was carried out with the following parameters: I = 0.25–0.5 A, V = 8 V, Q = 3.3 A h, anode loss = 93% of theoretical.

The reaction mixture was filtered and the residue was dried at 40°C/0.001 mmHg.

C$_{10}$H$_{14}$FeO$_4$: m.p. = 174°C. Found (%): Fe, 21.96; C, 47.20; H, 5.54. Calculated (%): Fe, 22.00; C, 47.95; H, 5.52.

The ferrous acetylacetonate crystallizing as yellowish-brown needles from absolute ethanol changed into ferric acetylacetone after being heated in acetylacetone with excess of oxygen.

When air or oxygen was bubbled through the electrolyte after the electrolysis was finished, it was possible isolate ferric acetylacetonate quantitatively. C$_{15}$H$_{21}$FeO$_6$: m.p. 182°C. Found (%): Fe, 15.73; C, 50.86; H, 6.25. Calculated (%): Fe, 15.82; C, 50.95; H, 5.95.

(g) Synthesis of nickel(II) acetylacetonate [64]

A solution of 12.9 g of LiClO$_4$ and 2.5 g of LiBr and including 75.4 of acetylacetone, in 100 cm^3 of dimethoxyethane, was electrolyzed between two nickel electrodes in an undivided cell. The experimental conditions for the electrosynthesis were: $I = 0.5$ A, $V = 15$ V, $Q = 5.65$ A h, anode loss = 100% of theoretical.

The deposit was filtered off, washed with DME and dried at $40°C/0.1$ mmHg. Yield: 36% of theoretical.

It was recommended to wash the crude product on the frit with water until it was no longer possible to detect any Br^- in the discharged washing water. The yield of green nickel(II) acetylacetonate then increased to 87%.

(h) Uranium β-diketonates [95]

The electrolysis of β-diketone solutions in ethanol ($0.4–0.6$ mol dm^{-3}) was carried out using an uranium anode and nickel cathode during 3–5 h at 45–60 mA either in an inert atmosphere (Ar, He) or in an oxidative one (O_2, dry air) with agitation. LiCl was used as supporting electrolyte (0.1 mol dm^{-3}). β-Diketones (acetylacetone (Hacac) and benzoylacetone (Hba)) were distilled before use.

The following products were isolated: (1) U(acac)$_4$ (grey, yield 60%) and U(ba)$_4$ (red–brown, 72%) (these solids were formed in an inert atmosphere); (2) UO$_2$(acac)$_2$·Hacac (orange, 59%) (the product was isolated after evaporation of ethanol, treatment of the solid formed with ether and its further evaporation); (3) UO$_2$(ba)$_2$·2Hba (red, structure **VII**, yield 30%, the same isolation procedure as for UO$_2$(acac)$_2$·Hacac). In the presence of H_2O_2 (30% aqueous solution), added in a ratio of 1:25 to the initial electrolyte, a 45% yield of UO$_2$(ba)$_2$·2Hba can be obtained after electrolysis and partial evaporation of the solvent [95].

(i) Dysprosium(III) acetylacetonate [95b]

The electrosynthesis of [Dy(acac)$_3$·EtOH] was carried out in an ethanolic solution of acetylacetone (12 mmol of Hacac in 30 cm^3 of EtOH) with a dysprosium cathode for 6 h. The current was 50 mA. LiCl (0.1 mol dm^{-3}) in ethanol was used as supporting electrolyte. Then the ethanol was evaporated, the solid was extracted with ether and the solution obtained was filtered and evaporated. The final product was white (almost quantitative yield, 99.2%).

(j) Cu + [(CH₃)₂NCS₂]₂ with simultaneous application of ultrasound [190]

The electrochemical synthesis was carried out according to the technique described by Tuck et al. [142]. The electrochemical cell was a 100-cm^3 tall beaker. The anode was a copper sheet (1 g); the cathode was a platinum foil. The electrolytic process in each case was applied for 2 h. n-Bu$_4$NBr (~0.05 g) was used as supporting electrolyte in the case of acetonitrile and EtOH–toluene solutions (100 cm^3). A current of 20 mA and agitation at 40 rev min^{-1} were applied in all the experiments. The isolated products were filtered, washed several times with small amounts of dry acetonitrile and dried in air. The TMU solutions were subjected to simultaneous

TABLE 2.7

Experimental conditions for the oxidative and electrochemical dissolution of copper in non-aqueous solutions of TMU

Experiment		TMU, g	Solvent	Use of ultrasound (ch) or (el)	Voltage,[‡] V (el)
(ch)*	(el)[†]				
A	I	1.25	AN	−	37
B	J	1.25	AN	+	33
C	K	1.25	EtOH + toluene	−	28
D	L	1.25	EtOH + toluene	+	31
E	M	1.30	DMF	−	35
F	N	1.30	DMF	+	33
G	O	1.30	DMSO	−	25
H	P	1.30	DMSO	+	29

*Chemical synthesis (oxidative dissolution).
†Electrochemical synthesis.
‡To produce an initial current of 20 mA.
In all experiments initial weight of Cu, $m_{Cu(init.)} = 1$ g; volume of solvent $= 100$ cm^3; agitation $= 40$ rev min^{-1}; $t = 2$ h.

ultrasonic treatment in some experiments (Tables 2.7 and 2.8) during the processes of synthesis using a weak source of ultrasound (ultrasonic cleaner Bransonic 12). Stronger sources of ultrasound have deliberately not been used, in order to prevent undesirably rapid metal dissolution, turbulent processes and superheating of the reaction zone.

2.3.3 Electrosynthesis of di- and polymetallic complexes

There are a few publications testifying that the electrosynthesis is a promising method of obtaining di- and polymetallic complexes.

The cadmium complex of composition CdL$_2$ [162], based on pyridine-2-thiols containing a trimethylsilyl substituent **XL** (LH) in the heteronucleous, was obtained under electrosynthesis conditions (AN, Et$_4$NClO$_4$). According to X-ray diffraction data, it represents a dimer (CdL$_2$)$_2$ **XLI**, consisting of two cadmium atoms in a trigonal-bipyramidal configuration [162], connected through a sulfide bridge. The possibility is not excluded that a nickel complex and, especially, a zinc complex with the ligand **XL** [162], have similar structures.

TABLE 2.8
Comparative results of the synthesis processes

Expt.	Solid		Cu(f/c),* %	$m_{Cu(diss.)}$,[†] g	$m_{Cu(diss.)}/m_{Cu(init.)}$ × 100%
	Color	M.p., °C			
A	Brown	262	20.31/20.91	0.040	4.0
B	Brown	265	20.42/20.91	0.123	12.3
C	Brown	260	19.98/20.91	0.048	4.8
D	Brown	263	20.17/20.91	0.149	14.9
E	Black	320	13.62/14.11	0.053	5.3
F	Black	323	13.55/14.11	0.138	13.8
G	Black	327	13.38/13.81	0.045	4.5
H	Black	325	13.40/13.81	0.150	13.0
I	Brown	260	20.17/20.91	0.062	6.2
J	Brown	262	20.03/20.91	0.180	18.0
K	Brown	264	20.14/20.91	0.070	7.0
L	Brown	261	19.90/20.91	0.171	17.1
M	Black	323	13.70/14.11	0.064	6.4
N	Black	325	13.59/14.11	0.175	17.5
O	Black	327	13.29/13.81	0.081	8.1
P	Black	327	13.35/13.81	0.195	19.5

*Analytical results: f, found; c, calculated.
[†]Weight of copper dissolved.

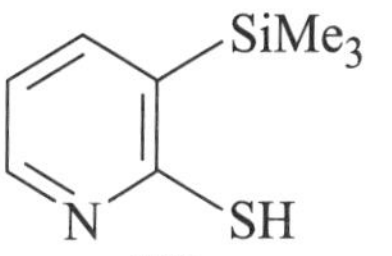

XL

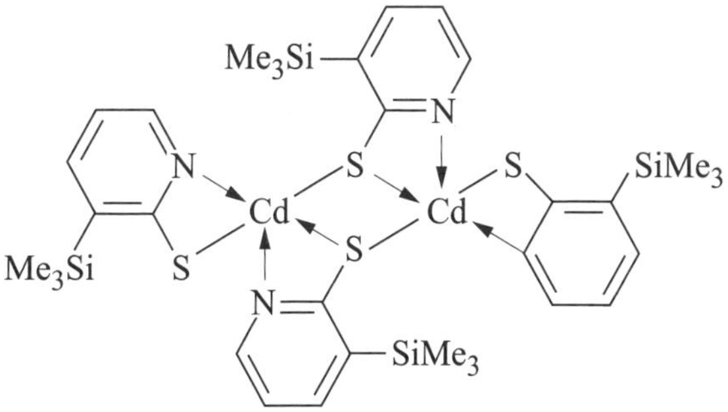

XLI

A series of binuclear complexes have been obtained by electrochemical dissolution of metals in the presence of diols [78] and thiols [77,163]. Thus, by electrochemical oxidation of copper in an acetonitrile solution of aromatic diols (dihydroxybenzene, tetrachloro-1,2-dihydroxybenzene and 2,2'-dihydroxybiphenyl) in the presence of N-(bipy) and P-(triphenylphosphine, bis-diphenylphosphinomethane)-donors, binuclear complexes of copper(I) were obtained [78]. The structure of one of them $(Cu_2[OC_6Cl_4(OH)_2]_2$ $[(C_6H_5)_2PCH_2P(C_6H_5)_2]_2)$ was determined by X-ray diffraction. The main feature of this structure is its eight-membered metal ring, containing two copper(I) atoms with two P–Cu–P bridges [78].

Under anodic dissolution of alkanethiols $R(SH)_2$, dimers are formed with the general formula M_2S_2R (where M = Tl [77], Cu and Ag [163]), in which a doubly deprotonated ligand is included. On the basis of these dimers, the series of adducts with the P-bases mentioned above was obtained, as well as with acetonitrile. X-ray diffraction data of one of them indicate both tetramerization and formation of octanuclear clusters with sulfide bridges, having the formula $[Cu_2S_2C_3H_7\text{-}1,2\text{-}(C_6H_5)_2PCH_2P(C_6H_5)_2]_4\cdot CH_3CN$ [163].

The series of di-, tri- and tetranuclear heteronuclear complexes $M[M'(CO)_n]_m$ (where M = Zn, Cd, In; M' = Co, Mn; m = 2, 3; n = 4, 5) and their adducts with bipyridyl and with N,N,N',N'-tetramethylenediamine, obtained under conditions of electrosynthesis starting from zero-valent metals and carbonyls $Co_2(CO)_8$ or $Mn_2(CO)_{10}$ in methanol, has been described [164] (see also Tables 2.5 and 2.6). The tetranuclear complex $[Cu_4(\mu\text{-}C_5H_{11})_4\cdot(Ph_2PCH_2PPh_2)_2]$ was synthesized through the interaction between copper, allylsulfide and bis(diphenylphosphine)methane under electrosynthesis conditions (AN, Et_4NClO_4) [165]. According to the X-ray diffraction data, a new type of eight-membered cyclic structure Cu_4S_4, in which the copper atoms are connected through sulfide bridges [165], is present in this complex.

Complexes with tetranuclear cations $[Et_3NH][M_4(SPh)_{10}]$, where M = Zn and Cd [166], are prepared by direct electrosynthesis with the participation of zinc or cadmium, thiophenol and triethylamine in acetonitrile (the supporting electrolyte is Et_4NClO_4). According to the X-ray diffraction data, a bicyclic structure with a terminal coordination of zinc and bridge SPh fragments [166] is present in this anion (M = Zn). The X-ray data showed that electrochemical synthesis makes it possible to create two types of hexanuclear cluster structures based on 4,6-dimethylpyrimidine-2-thione: an octahedral cluster with six Cu–Cu bonds and facet pyrimidinethione fragments is obtained for Cu_6L_6 [164,167]; a circular calixarene-like system with six cadmium atoms, bound by sulfide bridges, represents the second type of structure which was found for Cd_6L_6 [145b]. Taking into account the fact that the electrosynthesis of both the structures examined was carried out under similar conditions (AN, Et_4NClO_4), it is possible to suppose that the main factor in the creation of those structures is the possibility of both central atoms Cu^+, Cd^{2+} producing various polyhedra in their role as complex-builders.

The possibility of obtaining clusters based on electrochemically synthesized complexes is very attractive. Thus, the eight-membered cluster $[Cu_8(SC_5H_{11})_4(S_2CSC_5H_{11})_4]$ [168a] was synthesized by treatment with CS_2 of copper(I) amyl sulfide, obtained under electrosynthetic conditions in AN medium by using a copper anode. The interaction between the electrochemically obtained adduct of the cadmium complex of pyrimidine-2-thione with 1,10-phen (R = H, A = N, M = Co, L′ = phen, n = 2) and copper(I) hexafluoroacetylacetonate in AN medium leads to the formation of the tetraheteronuclear complex $[Cu_2(phen)(\mu_3-L)_2Cd_2 (nFacac)_2]\cdot 2AN$ [169], where L is the trianion of 1,3,5-tris(trifluormethyl)-1,3,5-trioxyhexane.

Additional data on bimetallic complexes obtained electrochemically are presented in Table 2.9.

2.3.4 Electrosynthesis of σ- and π-organometallic complexes

There are several references containing generalized data on the electrochemical synthesis of π-complexes involving transition metals [9,17,21,24]. They include the electrosynthesis of metal complexes of cycloolefins; for example, the copper complex of cyclooctadiene [24] $Cu(COD)_2ClO_4$ **XLII**, a nickel derivative of cyclooctatetraene [9] Ni(COT) **XLIII**, and bis(η^5-cyclopentadienyl) metallic derivatives of the type **XLIV** (M = Fe, Mn, Ni) [9,21,24].

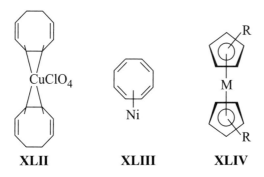

| **XLII** | **XLIII** | **XLIV** |

TABLE 2.9

Electrosynthesis of σ and π organometallic and polynuclear complexes

Metal	Initial system	Products	Ref.
Cd	RX, CH_3OH, benzene, $(n\text{-Pr})_4NX$; R = Me, Et, n-Bu,CF_3,Ph; X = Cl, Br, I	$[(n\text{-Pr})_4N][RCdX_2]$	38, 170, 171
	CH_3OH, RX, L (L = bipy, DMSO, phen, diox)	$RClX\cdot nL$	
Tl	RX, acetone	RTlX	14a
Sn	Et_2SO_4, EtI	$SnEt_4$	172
Sn	Benzene, methanol, RX (R = Me, Et; X = Cl, Br, I)	R_2SnX_2	173
	RX (R = Me, Et, n-Bu, Ph), DMSO, CH_3CN, diphos (1,2-bis(diphenylphosphino)ethane), bipy (L)	$R_2SnX_2\cdot L$	
Cd, Hg, Sn	Nonaqueous solvents, R_3SnCl	$Ph_3SnCdCl$ R_3SnSnR_3	174, 175
Mg	CH_3CN, ArX, bipy CH_3CN, $X(CH_2)_nX$ CH_3CN, R_4NX, ArX	$ArMgX\cdot bipy$ $MgX(CH_2)_nMgX\cdot 2bipy$ $R_4N[RMgX_2]\cdot CH_3CN$	14
Ti, Zr, Hf	CH_3CN or CH_3OH, RX (X = Cl, Br, I; R = Me, Et, Ph, $PhCH_2$) In the presence of bipy	$R_2MX_2\cdot nSolv$ $R_2MX_2\cdot bipy$	176
Ni, Pd	CH_3CN, RX (R = C_6F_5, $C_6H_5CH_2$; X = Cl, Br) C_6F_5Br, PEt_3	RMX $C_6F_5PdBr\cdot 2PEt_3$	177
	RCN (R = Me, Et, Ph, C_6F_5), PEt_3 $C_6H_5CH_2CN$, PEt_3	$RNiCN\cdot 2PEt_3$ $C_6H_5CH_2NiCN\cdot Et_3P$	
In	CH_2X_2/CH_3CN (X = Cl, Br, I)	X_2InCH_2X	71

TABLE 2.9   (contd.)

Metal	Initial system	Products	Ref.
Zn, Cd, In	Bipy, CH_3OH, C_6H_6, $Co_2(CO)_8$	$Zn[Co(CO)_4]_2 \cdot bipy$ $Cd[Co(CO)_4]_2$	164
	Bipy, CH_3OH, C_6H_6, $Mn_2(CO)_{10}$	$Zn[Mn(CO)_4]_2 \cdot bipy$ $Cd[Mn(CO)_5]_2 \cdot bipy$ $In[Mn(CO)_5]_3 \cdot bipy$	
Cd	CH_3OH, benzene, $Co_2(CO)_8$, L L = tmed, bipy)	$Cd[Co(CO_4)]_2 \cdot nL$	164
Ni, Co, Fe, Zn, Cu	CH_3CN, CpH or Me_5C_5H	$M(Cp)_2$ $M(Me_5C_5)_2$	178 96b
	CH_3CN, CpH, HX (X = Cl, Br), Bu_4NBr	$[Co(Cp)_2]_2CoX_4$	
Co, Cu, Au, Ag, Zn	CH_3CN, $Me_5C_5H_5$, THF, KPF_6,	$[Co(Me_5C_5)_2]_2(PF_6)$	178, 179
	CH_3CN, Me_5C_5H, Bu_4NBr	$[Co(Me_5C_5)_2]Br$	
	CH_3CN, $PhC{\equiv}CH$	$M(PhC{\equiv}C)_n$	
	CH_3CN, CpH, THF, L (L = PPh_3, phen)	$M(Cp) \cdot L$	
Pb anode	EtCl, EtMgCl	$PbEt_4$	21,
Pb cathode	EtBr	$PbEt_4$	45

diox, dioxane; tmed, N,N,N′,N′-tetramethylethylenediamine; CpH, cyclopentadiene.

Ferrocene **XLIV** (M = Fe, R = H) was obtained by electrochemical dissolution of an iron anode in a solution of cyclopentadienylthallium in DMF ([21a], p.135). The analogous manganese and nickel complexes have been synthesized as a result of the electrolysis of lithium, sodium or cadmium complexes of cyclopentadiene in a solution of THF ([21a] and references therein). However, the possibility of producing cyclopentadienides [180] and cyclopentadieniles [181–183] (see also Table 2.9) starting directly from cyclopentadiene itself, was later demonstrated. Thus, the cyclopentadienides of Mg, Zn and Al are formed by the electrolysis of cyclopentadiene solutions in DMSO, DMF or AN with anodes of the appropriate metal.

Metallocenes of the type **XLIV** (M = Fe, Co; R = H) have been obtained in THF or pyridine solutions in the presence of [Bu_4N]Br (see Table 2.9). The yields are 15–60% and depend on the nature of the metal, the solvent used and the temperature [178,183]. It was impossible to isolate nickelocene in these conditions. Cobaltocene and its derivatives have been obtained in the form of cobaltocene salts in a DMF solution with a yield of 20–30% [182].

In this respect, the electrosynthesis of metallocene complexes of the type **XLIV** starting from CH-acids RH (where R = c-C_5H_5, MeC_5H_4, 2,4-dimethyl-C_5H_3, C_5Me_5, indenyl and fluorenyl) are of especial interest [178]. It was shown that π-complexes R_2M (R = C_5H_5, 2,4-dimethyl-C_5H_3, M = Fe, Ni, Co, Zn; R = MeC_5H_4, fluorenyl, M = Fe, Ni; R = indenyl, M = Fe) could be obtained by electrosynthesis with anodes of ligands in AN, AN/1,2-dimethoxyethane, and AN/THF [178]. Moreover, the following mixed-ligand π-complexes have been synthesized: (C_5H_5)CuPPh$_3$, [Co(C_5H_5)$_2$]Br, {Co(C_5H_5)$_2$[CoCl$_4$]}, and {Co(C_5H_5)$_2$[CoBF$_4$]} [178].

In addition to these derivatives of cyclopentadiene, the mixed-ligand complexes of the types (π-Cp)$_n$M(CO)$_m$ and (π-Cp)$_n$M(CO)$_m$PR$_3$, together with several metal carbonyls (M = Fe, Co, Ni, Ti, V) have been synthesized on the basis of the same ligand system and CO under electrosynthesis conditions [168]. Returning again to the metal carbonyls, two details should be mentioned: (a) metal carbonyls could also be obtained as a result of reactions at a sacrificial cathode (Section 3.4) [14a]; (b) the mixed-ligand complexes $L_nM_m(CO)_{l-k}L'_k$ have been prepared from $L_nM_m(CO)_l$ by ligand substitution of CO in electrocatalysis conditions [21,184–186].

The σ-organometallic compounds have been obtained as a result of oxidative insertion of metals into C–H or C–X bonds (where X = Cl, Br, I) and are generalized in Tables 2.9 and 2.10. They can be also obtained by the electrochemical reduction of ketones using a sacrificial cathode (Section 2.4, Table 2.10). Detailed descriptions of their syntheses are presented in [14a,18].

The examples mentioned above testify to a limited application of the electrochemical method for the synthesis of metal π-complexes; this situation persists to the present.

TABLE 2.10
Data on the electrosynthesis of metal complexes at sacrificial cathodes

Metal	Initial system	Products	Ref.
Sn, Bi, Sb	$CH_2{=}CHCN$	$(NCCH_2CH_2)_nM$, $n = 2-4$	17
Hg	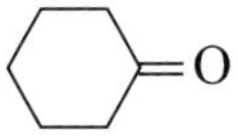 cyclohexanone ($=O$) / H_2SO_4 (5%)	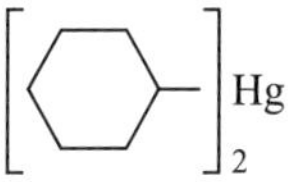 [cyclohexyl]$_2$Hg	187
Hg	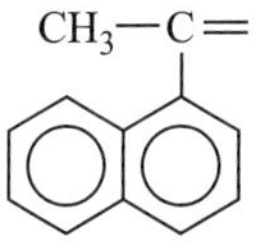 $CH_3{-}C{=}O$ (acetylnaphthalene) / H_2SO_4 (45%)	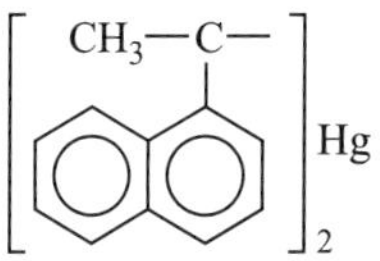 [$CH_3{-}C{-}$naphthyl]$_2$Hg	14a, 187
Hg	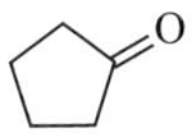 cyclopentanone (O) / $HCl + CH_3COOH$	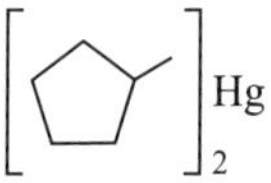 [cyclopentyl]$_2$Hg	187
Pb	$(C_2H_5)_2CO$, H_2SO_4	$[(C_2H_5)_2CH]_4Pb$	187
Hg	$RC_6H_4CH_2Br$, LiBr, CH_3OH	$Hg(CH_2C_6H_4R)_2$	14a, 17
Hg	$HL =$  (Ph, Ph, CH_3, I cyclopropene), CH_3CN, $(CH_3)_4NBr$	HgL_2	14a, 17
Sn	C_2H_5Br, CH_3CN, $(C_2H_5)_4NBr$	$Sn(C_2H_5)_4$	14a, 17
Pb	CH_3X (X = Cl, Br, I), C_2H_5X (X = Cl, Br), CH_3CN, $(C_2H_5)_4NBr$	PbR_4	14a, 188
Hg	$RHgBr$	$R_2Hg + HgBr_2$	14a
Hg	$Ph_2Pb(OAc)_2$	$R_2Hg + Pb + 2OAc^-$	14a
Hg	$(CH_3)_2CO$	$Hg[CH(CH_3)_2]$	14a
Hg	$Re_2(CO)_{10}$	$Hg[Re(CO)_5]_2$	14a
Hg	$Mn_2(CO)_{10}$	$Hg[Mn(CO)_5]_2$	14a

Experimental Procedures

(a) Electrochemical preparation of bimetallic adducts [Cr$_2$(DMSO)$_{20}$][S$_2$O$_7$]$_3$ and [Fe$_2$(DMSO)$_{10}$][S$_2$O$_7$]$_3$·SO$_2$ [81]

The metals Cr and Fe were dissolved electrolytically in DMSO–SO$_2$ to form crystalline solvated metal disulfates, which are the only binary systems capable of oxidizing sulfur(IV) to sulphur(VI), in the form of the disulfate ion.

All processes were carried out under a dry oxygen-free N$_2$ atmosphere. Solvents were dried over molecular sieve 4A, deoxygenated, and redistilled, and then saturated with sulfur dioxide (dried over phosphorus(V) oxide and concentrated sulfuric acid).

A platinum wire was used as cathode and 0.5 mol dm^{-3} Et$_4$N(ClO$_4$) (or 0.1 mol dm^{-3} Et$_4$N[BF$_4$]) as supporting electrolyte.

The rate of formation of these complexes depends on the potential difference of the electrolytic cell, and on the current.

In the case of iron, the choice of the supporting electrolyte plays an important role. Thus, use of Et$_4$N[BF$_4$] always formed a mixture of sulfate and disulfate, whereas use of Et$_4$N[ClO$_4$] always gave disulfate. Generally, a potential difference of 10 V was applied across the cell, and currents of 40–100 mA were obtained.

The thermal decomposition of the solvated disulfates has been studied up to 1000°C. The complex [Cr$_2$(DMSO)$_{20}$][S$_2$O$_7$]$_3$ began to lose DMSO at 110°C and at 400°C Cr$_2$(S$_2$O$_7$)$_3$ was obtained. Further heating resulted in loss of SO$_3$ to give Cr$_2$O$_3$ at 700°C. From Fe$_2$(DMSO)$_{10}$[S$_2$O$_7$]$_3$·SO$_2$, at 290°C Fe(S$_2$O$_7$)$_3$ was obtained. Further heating resulted in loss of SO$_3$ to give Fe$_2$(SO$_4$)$_3$ at 395°C, followed by loss of more SO$_3$ to give Fe$_2$O$_3$ at 900°C.

(b) Electrochemical synthesis of metallocenes and polymetallocenes [178]

A stout platinum sheet (80 mm × 4 mm) formed the cathode; metals (Fe, Co, Ni, Cu, Zn, Ag, Au) were cleaned with 5 mol dm^{-3} HCl, then polished with fine emery cloth. Some reagents were used as supplied and others were suitably treated.

The electrolyte (ca 200 mg) was added to a 200-cm^3 round-bottom Schlenk flask into which THF and acetonitrile (typically in a 3:1 ratio) were distilled under Ar. Monomeric cyclopentadiene (ca 4 cm^3) was then condensed into it (similarly for methylcyclopentadiene). The flask was fitted with a tight rubber septum and maintained at −80°C.

Alternatively, 4 cm^3 of Me$_5$C$_5$H or 2,4-dimethylpentadiene was added to the solvent system at room temperature.

The electrolytic cell was a 200-cm^3 tall beaker with a bottom outlet. The process was carried out at 40–50 V at room temperature; a voltage of 50 V produced an initial current of 15–30 mA.

Electrolysis over 1–3 h dissolved approximately 100 mg of metal. In the case of metallocenes, the organometallic was extracted with hexane, the solvent removed and the complex sublimed in vacuo at $80°C/13$ Pa. In most cases solutions eventually turned brown.

Complexes precipitated in each electrolytic process were collected, washed with ether, and then dried in vacuo or over P_4O_{10}.

Among the complexes obtained from different solution compositions are the following:

Fe(Cp)$_2$ $I_{in} = 150$ mA, $t = 50$ min, anode loss $= 100$ mg, yield 70%
Fe(MeCp)$_2$ $I_{in} = 15$ mA, $t = 55$ min, anode loss $= 12$ mg, yield 66%
Ni(MeCp)$_2$ $I_{in} = 24$ mA, $t = 50$ min, anode loss $= 25$ mg, yield 90%
Ni(Cp)$_2$ $I_{in} = 12$ mA, $t = 135$ min, anode loss $= 29$ mg, yield 50%

2.4 ELECTROSYNTHESIS AT SACRIFICIAL CATHODES

In some cases the electrolysis of ketones, RHal or unsaturated compounds can lead to the formation of the organometallic compounds with metallic cathode dissolution. Only certain metals can dissolve by this method, mainly Hg, Pb and Sn [14a]. Konev et al. [17] noted that the phenomenon of metal cathode dissolution is paradoxical. They suppose that free radicals possibly take part in these processes: organic radicals are formed on the cathode surface and then react with the metal. Cathode dissolution reactions are summarized in Table 2.10 [14a,17,187,188]. It should be noted that at present this method has a limited applicability (the majority of the synthesized complexes have been obtained by anodic dissolution of metals).

2.5 CONCLUSIONS

The summarized research results testify to the wide application of electrosynthesis as a method of obtaining molecular complexes and metal chelates. Various metals (of almost all groups of the Periodic Table) and ligands, including both inorganic and organic compounds, are used under the conditions of electrochemical synthesis.

At the same time, representation of the elements of the secondary subgroups of groups IV–VII and the actinides is highly limited among the metals. Any information about the possibility of using lanthanides (except cerium) in the electrosynthesis of metal complexes is almost absent in the literature. If this area is developed successfully, it should be possible to discover not only a route to new adducts of lanthanide salts with N-,P-,O-,S-bases, but also to metal chelates of rare earth elements, a branch of synthetic chemistry on which little work has been done.

The prospects of applying electrosynthesis in order to obtain π-complexes and chelates with macrocyclic ligands [199] are not clear, although successful electrosyntheses of some metal complexes of these types (see Sections 2.3.2 and 2.3.4) is very promising. Undoubtedly, the lanthanides should be more widely used in future electrosynthetic procedures. The electrosynthesis of peroxo complexes ([18] and references therein) seems very interesting because of the possibility of obtaining metals with unusually high oxidation numbers, and should also be developed. Moreover, the combination of the electrochemical dissolution of metals with simultaneous ultrasonic treatment of the reaction system [25e,124,156,189,190] could help to avoid the typical problems of electrochemical procedures and to increase the yields. In this respect, the sonoelectrochemical cell reported by Compton et al. [25e] can be considered as the first attempt at such a combination. Several electrochemical reactions of metal complexes obtained on sacrificial cathodes are described in the literature [14a,17,21,24]. However, except for the studies reviewed in these publications, there is no information in the papers published in recent years about use of a sacrificial cathode as a source of metal.

The material included in this chapter testifies to the importance of electrochemical dissolution of metal as a method of synthesizing coordination compounds. The recent papers in this area [1,3,14b,18,21b,29–31,70,104–115,124,144,145,156–160,189–199] strongly support this idea.

ACKNOWLEDGMENTS

The authors are very grateful to Professor Antonio Sousa (University of Santiago-de-Compostela, Spain) and Professor Dennis G. Tuck (University of Windsor, Canada) for the opportunity to consult reprints of their publications, and to Professor Jim D. Atwood (State University of New York at Buffalo) for the critical revision of this chapter.

REFERENCES

1. *Direct Synthesis of Coordination Compounds* (Ed.: Skopenko, V.V.). Kiev: Ventury (1997), pp. 109–128.
2. Garnovskii, A.D.; Ryabukhin, Yu.I.; Kuzharov, A.S. *Koord. Khim.* **10**(8), 1011 (1984).
3. Garnovskii, A.D.; Kharisov, B.I.; Gójon-Zorrilla, G.; Garnovskii, D.A. *Russ. Chem. Rev.* **64**(3), 201 (1995).
4. Klabunde, K.J. *Chemistry of Free Atoms and Particles.* New York: Academic Press (1980).
5. Lavrentiev, I.P.; Khidekel, M.L. *Russ. Chem. Rev.* **52**(4), 337 (1983).
6. Gerdes, B. *J. Pract. Chem.* **26**(11), 257 (1882).
7. Scillard, B. *Z. Electrochem.* **12**(22), 393 (1906).
8. Zamyatin, V.M.; Kukushkin, Yu.N.; Makarenya, A.A. *Lev Alexandrovich Chugaev.* Nauka: Moscow (1973), pp. 73, 102.
9. Lehmkuhl, H. *Synthesis* **7**, 377 (1973).
10. Bogdashev, N.N.; Garnovskii, A.D.; Grigor'ev, V.P.; Osipov, O.A. *Izvest. Severo-Kavkaz. Nauk Tsentra Visshei Shkoli.* **2**, 28 (1973).
11. Osipov, O.A.; Bogdashev, N.N.; Grigoriev, V.P. Garnovskii, A.D. USSR Patent 401674 (1973).
12. Osipov, O.A.; Bogdashev, N.N.; Grigoriev, V.P.; Garnovskii, A.D.; Gontmakher, N.M. USSR Patent 485115 (1975).
13. Bogdashev, N.N.; Garnovskii, A.D.; Osipov, O.A.; Grigoriev, V.P.; Gontmakher, N.M. *Zh. Obshch. Khim.* **46**(3), 675 (1975).
14. a) Tuck, D.G. *Pure Appl. Chem.* **51**(9), 2005 (1979); b) Tuck, D.G. In *Molecular Electrochemistry of Inorganic, Bioinorganic and Organometallic Compounds.* Dordrecht: Kluwer (1993), pp. 15–31.
15. Castro, J.A.; Romero, J.; García-Vázquez, J.A.; Macías, A.; Sousa, A.; Englert, U. *Polyhedron* **12**, 1391 (1993).
16. Laube, B.L.; Schulbach, C.D. *Prog. Inorg. Chem.* **14**, 65 (1971).
17. Konev, V.A.; Kukushkin, V.Yu.; Kukushkin, Yu.N. *Zh. Neorgan. Khim.* **31**(6), 1466 (1986).
18. Chakravorti, M.C.; Subrahmanyam, G.V.P. *Coord. Chem. Rev.* **135/136**(1), 65 (1994).
19. Abers, M.O.; Coville, N.J. *Coord. Chem. Rev.* **53**, 227 (1984).
20. Thaylor, M.; Tuck, D.G. *Inorg. Synth.* **22**, 135 (1983).
21. a) Kukushkin, V.Yu.; Kukushkin, Yu.N. *Theory and Practice of Coordination Compounds Synthesis.* Leningrad: Nauka (1990), pp. 127–141; b) Davies, J.A.; Hockensmith, C.M.; Kukushkin, V.Yu.; Kukushkin, Yu.N. *Synthetic Coordination Chemistry. Theory and Practice.* Singapore: World Scientific Publishing (1996), Chap. 7.
22. *Electrochemical Synthesis* (Eds: Little, R.D.; Weinberg, N.L.). New York: Marcel Dekker (1991).
23. Lehmkuhl, H. In *Electrochemistry of Organic Compounds* (Ed.: Baizer, M.). Moscow: Mir (1976), p. 442.
24. Tomilov, A.P.; Chernih, I.N.; Kargin, Yu.M. *Electrochemistry of Elementoorganic*

Compounds. Elements of I, II, III groups of Periodic Table and Transition Metals. Moscow: Nauka (1985).

25. a) *Ultrasound: its Chemical, Physical, and Biological Effects* (Ed.: Suslick, K.S.). Weinheim: VCH (1988); b) Mason, T.J. *Advances in Sonochemistry*, Vol. 1. London: JAI Press (1990); c) Cintas, P. *Activated Metals in Organic Synthesis*. Boca Raton: CRC Press (1993), pp. 61–70; d) Luche, J. L.; Cintas, P. Ultrasound-induced activation of metals: principles and applications in organic synthesis. In: *Active Metals: Preparation, Characterization, Applications* (Ed.: Fuerstner, A.). Weinheim: VCH (1996), pp. 133–190; (e) Compton, R.G.; Eklund, J.C.; Page, S.D.; Rebbit, T.O. *J. Chem. Soc., Dalton Trans.* 389 (1995).

26. Grobe, J.; Keil, M.; Schneider, B.; Zimmermann, H. *Z. Naturforsch. Teil B* **35**, 428 (1980).

27. Kumar, N.; Tuck, D.G. *Can. J. Chem.* **60**, 2579 (1982).

28. Kumar, N.; Tuck, D.G. *Inorg. Chim. Acta* **95**(2), 211 (1984).

29. McAuliffe, C.A.; Pritchard, R.G.; Bermejo, M.R.; García-Vázquez, A.; Macías, A.; Sanmartín, J.; Romero, J.; Sousa, A. *Acta Crystallogr., Sect. C* **48**, 1316 (1992).

30. McAuliffe, C.A.; Pritchard, R.G.; Bermejo, M.R.; García-Vázquez, A.; Macías, A.; Sanmartín, J.; Romero, J.; Sousa, A. *Acta Crystallogr., Sect. C* **48**, 1841 (1992).

31. Labisbal, E.; Romero, J.; García-Vázquez, J.A.; Sousa, A.; Castiñeiras, A. J. *Crystallogr. Spect. Res.* **23**(11), 895 (1993).

32. Habeeb, J.J.; Said, F.F.; Tuck, D.G. *Inorg. Nucl. Chem. Lett.* **15**(2), 113 (1979).

33. Said, F.F.; Tuck, D.G. *Can. J. Chem.* **59**, 62 (1981).

34. Khan, M.; Colin, O.; Tuck, D.G. *Can. J. Chem.* **59**, 2714 (1981).

35. Chandha, R.K.; Kumar, R.; Tuck, D.G. *Can. J. Chem.* **65**, 1336 (1987).

36. Kumar, R.; Tuck, D.G. *Inorg. Chim. Acta* **157**(1), 51 (1989).

37. Kumar, R.; Tuck, D.G. *Can. J. Chem.* **67**, 127 (1989).

38. Habeeb, J.J.; Osman, A.; Tuck, D.G. *J. Chem. Soc., Chem. Commun.* 379 (1976).

39. Habeeb, J.J.; Neilson, L.; Tuck, D.G. *Inorg. Chem.* **17**(2), 306 (1978).

40. Said, F.F.; Tuck, D.G. *Inorg. Chim. Acta* **59**(1), 1 (1982).

41. Hayes, P.C.; Osman, A.; Seudeal, N.; Tuck, D.G. *J. Organomet. Chem.* **291**(1), 1 (1985).

42. Habeeb, J.J.; Tuck, D.G. *J. Chem. Soc., Chem. Commun.* 808 (1975).

43. Habeeb, J.J.; Said, F.F.; Tuck, D.G. *J. Chem. Soc., Dalton Trans.* 1161 (1980).

44. Habeeb, J.J.; Said, F.F.; Tuck, D.G. *J. Chem. Soc., Dalton Trans.* **1**, 118 (1981).

45. Kumar, R.H.; Mabrouk, H.E.; Tuck, D.G. *J. Chem. Soc., Dalton Trans.* 1045 (1988).

46. Green, J.H.; Kumar, R.; Seudeal, N.; Tuck, D.G. *Inorg. Chem.* **28**(1), 123 (1989).

47. Habeeb, J.J.; Said, F.F.; Tuck, D.G. *Can. J. Chem.* **55**, 3883 (1977).

48. Christofis, O.; Habeeb, J.J.; Sievensz, R.S.; Tuck, D.G. *Can. J. Chem.* **56**, 2269 (1978).

49. Hencher, J.L.; Khan, M.; Said, F.F.; Sieler, R.; Tuck, D.G. *Inorg. Chem.* **21**(7), 2787 (1982).

50. Habeeb, J.J.; Neilson, L.; Tuck, D.G. *Can. J. Chem.* **55**, 2631 (1977).

51. Oldham, C.; Tuck, D.G. *J. Chem. Educ.* **59**(5), 420 (1982).

52. Chadha, R.K.; Kumar, R.; López-Grado, J.R.; Tuck, D.G. *Can. J. Chem.* **66**, 2151 (1988).

53. Kumar, N.; Tuck, D.G. *Inorg. Chem.* **22**, 1951 (1983).

54. Kumar, N.; Tuck, D.G. *Can. J. Chem.* **62**, 1701 (1984).

55. Sogo, T.; Romero, J.; Sousa, A.; De Blas, A.; Durán, M.L. *Z. Naturforsch. Teil B* **43**, 611 (1988).

56. *Chemistry of Pseudohalides* (Eds.: Golub, A.M.; Coler, H.; Skopenko, V.V.). Amsterdam: Elsevier (1986).

57. Garnovskii, A.D. Sadimenko, A.P.; Osipov, O.A.; Tsintsadze, G.V. *Hard–Soft Interactions in the Coordination Chemistry.* Rostov-on-Don: Rostov State University Press (1986).

58. Burmeister, J.L. *Coord. Chem. Rev.* **105**, 77 (1990).

59. Shreider, V.A. *Izv. Akad. Nauk SSSR, Ser. Khim.* **8**, 1832 (1984).

60. Kucheiko, S.I.; Kessler, V.G.; Turova, N.Ya. *Koord. Khim.* **13**(8), 1043 (1987).

61. Pisarevskii, A.P.; Martinenko, L.I. *Koord. Khim.* **20**(5), 324 (1994).

62. Kalinnikov, V.T.; Rakitin, Yu.V. *Introduction to Magnetochemistry. Magnetic Methods in Chemistry.* Moscow: Nauka (1980).

63. a) Travis, W.; Park, J.; Garza, G.; Ross, C.B.; Smith, D.M.; Croks, R.M. *Mater. Res. Soc. Symp. Proc.* **271**, 857 (1992); b) Chakravorti, M.C.; Subrahmanyam, G.V.B. *Polyhedron* **11**(24), 3191 (1992).

64. Eisenbach, W.; Lehmkuhl, H.; Wilke, G. Canadian Patent 1024466 (1978).

65. a) Kessler, V.G.; Shevelkov, A.V.; Khvorih, G.V.; Seisenbaeva, G.A.; Turova, N.Ya.; Drobot, D.V. *Zh. Neorg. Khim.* **40**(9), 1477 (1995); b) Kaabak, L.V.; Tomilov, A.P. *Zh. Obshch. Khim.* **67**(2), 341 (1997).

66. Banait, J.S.; Pahil, P.K. *Bull. Electrochem.* **5**(4), 264 (1989).

67. Mabrouk, H.E.; Tuck, D.G. *Can. J. Chem.* **67**, 746 (1989).

68. Mabrouk, H.E.; Tuck, D.G. *J. Chem. Soc., Dalton Trans.* 2539 (1988).

69. Chakravorti, M.C.; Subrahmanyam, G.V.B. *Can. J. Chem.* **70**, 836 (1992).

70. Castro, R.; García-Vázquez, J.A.; Romero, J.; Sousa, A.; Prichard, R.; *Polyhedron* **12**, 2241 (1993).

71. Annan, T.A.; Tuck, D.G.; Khan, M.A.; Peppe, C. *Organometallics* **10**(7), 2159 (1991).

72. Habeeb, J.J.; Tuck, D.G. *Inorg. Synth.* **19**, 257 (1979).

73. Habeeb, J.J.; Neilson, L.; Tuck, D.G. *Synth. React. Inorg. Metal-Org. Chem.* **6**(2), 105 (1976).

74. Habeeb, J.J.; Tuck, D.G. *J. Chem. Soc., Chem. Commun.* 600 (1975).

75. Shreider, V.A.; Volpin, I.M.; Gorbunova, Yu.E. *Izv. Akad. Nauk SSSR, Ser. Khim.* **5**, 958 (1988).

76. Mabrouk, H.E.; Tuck, D.G. *Inorg. Chim. Acta* **145**(3), 237 (1988).

77. Geloso, C.; Mabrouk, H.E.; Tuck, D.G. *J. Chem. Soc., Dalton Trans.* 1759 (1989).

78. Annan, T.A.; Kickman, R.; Tuck, D.G. *Can. J. Chem.* **69**(2), 251 (1991).

79. Annan, T.A.; Kumar, R.; Tuck, D.G. *J. Chem. Soc., Dalton Trans.* 11 (1991).

80. a) Kumar, N.; Tuck, D.G.; Watson, K.D. *Can. J. Chem.* **65**, 740 (1987); b) El-Asmy, A.A.; Saad, E.M.; El-Shahawi, M. *Transition Met. Chem.* **19**, 406 (1994).

81. Graham, N.K.; Gill, J.B.; Goodall, G.C. *J. Chem. Soc., Dalton Trans.* 1363 (1983).

82. Graham, N.K.; Gill, J.B.; Goodall, G.C. *Aust. J. Chem.* **36**, 1991 (1983).

83. a) Tallon, J.; García-Vázquez, J.A.; Romero, J.; Louro, M.S.; Sousa, A.; Chen, Q.;

Chang, Y.; Zubieta, J. *Polyhedron* **14**(17–18), 2309 (1995); b) Hencker, J.L.; Khan, M.; Said, F.F.; Seiler, R.; Tuck, D.G. *Inorg. Chem.* **21**, 2787 (1982).

84. Kumar, R.; Tuck, D.G. *Inorg. Chem.* **29**, 1444 (1979).

85. Castro, J.A.; Romero, J.; García-Vázquez, J.A.; Castiñeiras, A.; Sousa, A.; Zubieta, J. *Polyhedron* **15**(20–21), 2841 (1996).

86. a) Chakravorti, M.C.; Ganguly, S.; Subrahmanyam, G.V.B.; Bhattacharjee, M. *Polyhedron* **12**(6), 683 (1993); b) Chakravorti, M.C.; Ganguly, S.; Bhattacharjee, M. *Polyhedron* **12**(1), 55 (1993);

87. Shirokii, V.L.; Knizhnikov, V.A.; Ryabtsev, A.N.; Dikusar, E.A.; Bazhanov, A.V.; Mayer, N.A. *Zh. Obshch. Khim.* **67**(7), 1120 (1997).

88. Garnovskii, A.D. *Koord. Khim.* **18**(7), 675 (1992).

89. Garnovskii, A.D.; Nivorozhkin, A.L.; Minkin, V.I. *Coord. Chem. Rev.* **126**(1–2), 1 (1993).

90. Lehmkuhl, H.; Eisenbach, W.Ju. *Liebigs Ann. Chem.* **4**, 672 (1975).

91. Habeeb, J.J.; Tuck, D.G.; Walters, F.H. *J. Coord. Chem.* **8**(1), 27 (1978).

92. Bustos, L.; Green, J.H.; Hencher, J.L.; Khan, M.A.; Tuck, D.G. *Can. J. Chem.* **61**, 2141 (1983).

93. Matassa, L.; Kumar, N.; Tuck, D.G. *Inorg. Chim. Acta* **109**(1), 19 (1985).

94. Mazurenko, E.A. Dr. Hab. Thesis. Kiev: IONH Akad. Nauk. Ukr. SSR. (1987).

95. Kostyuk, N.N.; Kolevich, T.A.; Shirokii, V.L.; Umreiko, D.S. *Koord. Khim.* **15**(12), 1704 (1989); b) Kostyuk, N.N.; Shirokii, V.L.; Vinokurov, I.I.; Mayer, N.A. *Zh. Obshch. Khim.* **64**(9), 1432 (1994).

96. a) Kostyuk, N.N.; Shirokii, V.L.; Dik, T.A.; Vinokurov, I.I.; Umreiko, D.S. *Koord. Khim.* **17**(11), 1573 (1991); b) Kostyuk, N.N.; Dik, T.A.; Trebnikov, A.G.; Shirokii, V.L. *Proc. News of Electrochemistry of Organic Compounds*, Novosibirsk (1998), pp. 69–72.

97. Kostyuk, N.N.; Shirokii, V.L.; Vinokurov, I.I.; Dik, T.A.; Umreiko, D.S.; Erdman, A.A. *Zh. Neorg. Khim.* **37**(1), 68 (1992).

98. Bogdashev, N.N. Ph.D. Thesis. Rostov-on-Don State University (1975).

99. Alvarino, C.; Romero, J.; Sousa, A.; Durán, M.L. *Z. Anorg. Allg. Chem.* **556**, 223 (1988).

100. Strerton, N.M.; Fenton, D.E.; Hewson, G.J.; McLean, C.H.; Bastida, R.; Romero, J.; Sousa, A.; Castellano, E.E. *J. Chem. Soc., Dalton Trans.* 1059 (1988).

101. Castiñeiras, A.; Castro, J.A.; Durán, M.L.; García-Vázquez, J.A.; Macías, A.; Romero, J.; Sousa, A. *Polyhedron* **8**(21), 2543 (1989).

102. Durán, M.L.; García-Vázquez, J.A.; Macías, A.; Romero, J.; Sousa, A. *Z. Anorg. Allg. Chem.* **573**, 215 (1986).

103. Bastida, R.; Bermejo, M.R.; Louro, M.A.; Romero, J.; Sousa, A.; Fenton, D.E. *Inorg. Chim. Acta* **145**(21), 167 (1988).

104. Castro, J.; Romero, A.; García-Vázquez, J.A.; Durán, M.L.; Castiñeiras, A.; Sousa, A.; Fenton, D.E. *J. Chem. Soc., Dalton Trans.* 3255 (1990).

105. Labisbal, E.; Romero, J.; De Blas, A.; García-Vázquez, J.A.; Durán, M.L.; Castiñeiras, A.; Sousa, A.; Fenton, D.E. *Polyhedron* **11**(1), 53 (1992).

106. Labisbal, E.; De Blas, A.; García-Vázquez, J.A.; Romero, J.; Durán, M.L.; Sousa, A.; Bailey, N.A.; Fenton, D.E.; Leeson, P.B.; Parish, R.V. *Polyhedron* **11**(2), 227 (1992).

107. Gaber, M.; Mabrouk, H.E.; Ba-Issa Abdalls, A.; Ayad, M.M. *Monatsh. Chem.* **123**, 1089 (1992).

108. Labisbal, E.; García-Vázquez, J.A.; Macías, A.; Romero, J.; Sousa, A.; Englett, U.; Fenton, D.E. *Inorg. Chim. Acta* **203**(1), 67 (1993).

109. McAuliffe, C.A.; Pritchard, R.G.; Lauces, L.; García-Vázquez, J.A.; Romero, J.; Bermejo, M.R.; Sousa, A. *Acta Crystallogr., Sect. C* **49**, 587 (1993).

110. Labisbal, E.; Romero, J.; Durán, M.L.; García-Vázquez, J.A.; Sousa, A.; Russo, U.; Pritchard, R.; Renson, M. *J. Chem. Soc., Dalton Trans.* 755 (1993).

111. Labisbal, E.; Romero, J.; García-Vázquez, J.A.; Gómez, C.; Sousa, A. *Polyhedron* **13**(11), 1735 (1994).

112. Labisbal, E.; García-Vázquez, J.A.; Romero, J.; Sousa, A.; Castiñeiras, A.; Maichle, C.; Russo, U. *Inorg. Chim. Acta* **223**(1), 87 (1994).

113. Labisbal, E.; García-Vázquez, J.A.; Romero, J.; Picos, S.; Sousa, A.; Castiñeiras, A.; Maichle-Mössmer, C. *Polyhedron* **14**(5), 663 (1995).

114. Bastida, R.; De Blas, A.; Fenton, D.E.; Rodríguez, T. *Polyhedron* **11**(21), 2739 (1992).

115. Bastida, R.; De Blas, A.; Fenton, D.E.; Rodríguez, T. *Inorg. Chim. Acta* **203**(1), 47 (1993).

116. Garnovskii, A.D. *Russ. J. Coord. Chem.* **6**(5), 368 (1993).

117. Mistryukov, A.E.; Vasil'chenko, I.S.; Sergienko, V.S.; Nivorozhkin, L.E.; Kochin, S.G.; Porai-Koshitz, M.A.; Nivorozhkin, L.E.; Garnovskii, A.D. *Mendeleev Commun.* **1**, 30 (1992).

118. Antsishkina, A.C.; Porai-Koshitz, M.A.; Vasil'chenko, I.S.; Nivorozhkin, A.L.; Garnovskii, A.D. *Dokl. Akad. Nauk SSSR* **330**(1), 54 (1993).

119. Garnovskii, A.D.; Antsishkina, A.S.; Vasil'chenko, I.S.; Sergienko, V.S.; Kochin, S.G.; Nivorozhkin, A.L.; Mistryukov, A.E.; Uraev, A.I.; Garnovskii, D.A.; Sadikov, G.G.; Porai-Koshits, M.A. *Zh. Neorg. Khim.* **40**(1), 67 (1995).

120. Garnovskii, A.D.; Burlov, A.D.; Yusman, T.A.; Zaletov, V.G. *Koord. Khim.* **21**(1), 62 (1995).

121. Chikina, N.L.; Abakarov, G.M.; Shneider, A.A.; Kuren, S.G.; Litvinov, V.V.; Zaletov, V.G.; Garnovskii, A.D.; Sadekov, I.D. *Zh. Obshch. Khim.* **58**(11), 2496 (1988).

122. Sadekov, I.D.; Maksimenko, A.A.; Maslakov, A.G.; Minkin, V.I. *J. Organomet. Chem.* **391**(2), 179 (1990).

123. a) Garnovskii, D.A.; Sousa, A.; Sigeikin, S.G.; Vasilchenko, I.S.; Kurbatov, V.P.; Garnovskii, A.D. *Zh. Obshch. Khim.* **66**(1), 143 (1996); b) Garnovskii, D.A.; Burlov, A.S.; Garnovskii, A.D.; Vasilchenko, I.S.; Sousa, A. *Zh. Obshch. Khim.* **66**(9), 1546 (1996).

124. a) Kharisov, B.I.; Blanco, L.M.; Garnovskii, A.D. et al. *Polyhedron* **17**(2–3), 381 (1998); b) Kharisov, B.I.; Blanco, L.M.; Garnovskii, A.D. et al. *Polyhderon*, in press (1999).

125. Castro, J.A.; Vilasandes, J.E.; Romero, J.; García-Vázquez, J.A.; Durán, M.L.; Sousa, A.; Castellano, F.E.; Zukerman-Schpector, J. *Z. Anorg. Allg. Chem.* **612**(1), 83 (1992).

126. Castro, J.A.; Romero, J.; García-Vázquez, J.A.; Castiñeiras, A.; Durán, M.L.; Sousa, A. *Z. Anorg. Allg. Chem.* **645**, 155 (1992).

127. Castro, J.A.; Romero, J.; García-Vázquez, J.A.; Durán, M.L.; Sousa, A.; Castellano, E.E.; Zukerman-Schpector, J. *Polyhedron* **11**(2), 235 (1992).

128. Castro, J.A.; Romero, J.; García-Vázquez, J.A.; Sousa, A.; Castellano, E.E.; Zukerman-Schpector, J. *Polyhedron* **12**(1), 31 (1993).

129. Castro, J.A.; Romero, J.; García-Vázquez, J.A.; Castiñeiras A.; Sousa A. *Z. Anorg. Allg. Chem.* **619**, 601 (1993).

130. Castro J.A.; Romero J.; García-Vázquez J.A.; Sousa, A.; Castellano, E.E.; Zukerman-Schpector, J. *J. Coord. Chem.* **28**, 125 (1993).

131. Castro, J.A.; Romero, J.; García-Vázquez, J.A.; Sousa, A.; Castellano, E.E.; Zukerman-Schpector, J. *J. Coord. Chem.* **30**, 165 (1993).

132. Castro, J.A.; Romero, J.; García-Vázquez, J.A.; Macías, A.; Sousa, A.; Englert, U. *Acta Crystallogr., Sect. C* **50**, 369 (1994).

133. Castro, J.A.; Romero, J.; García-Vázquez, J.A.; Castiñeiras, A.; Sousa, A. *Transition Met. Chem.* **19**, 343 (1994).

134. Henning, H. *Z. Chem.* **11**, 81 (1971).

135. a) Panova, G.V.; Vikulova, N.K.; Potapov, V.M. *Usp. Khim.* **49**, 1234 (1980); b) Garnovskii, A.D.; Alekseenko, V.A.; Burlov, A.S.; Nedzvetskii, V.S. *Zh. Neorg. Khim.* **36**(4), 886 (1991).

136. Ivanskii, V.I. *Chemistry of Heterocyclic Compounds.* Moscow: Vissh. Shkola (1978).

137. a) Annan, T.A.; Peppe, C.; Tuck, D.G. *Can. J. Chem.* **68**, 423 (1990); b) Annan, T.A.; Peppe, C.; Tuck, D.G. *Can. J. Chem.* **68**, 1598 (1990).

138. a) Sanmartín, J.; Bermejo, M.R.; Romero, J.; García-Vázquez, J.A.; Sousa, A.; Brodbeck, A.; Castiñeiras, A.; Hiller, W.; Strahle, J. *Z. Naturforsch. Teil B* **48**, 431 (1993); b) Mabrouk, H.E.; El-Asmy, A.A.; Al-Ansi, T.Y.; Mounir, M. *Bull. Soc. Chim. Fr.* **128**(2), 309 (1991).

139. Svyatkina, L.I.; Baikalova, L.V.; Domnina, E.S. *Koord. Khim.* **21**(6), 496 (1995).

140. Gontmakher, N.M.; Grigoriev, V.P.; Abakumov, G.A.; Okhlobistin, O.Yu; Nechaeva, O.N.; Klimov, E.S.; Zaletov, V.G. *Dokl. Akad. Nauk SSSR* **228**(4), 346 (1976).

141. a) Mabrouk, H.E.; Tuck, D.G.; Khan, M.A. *Inorg. Chim. Acta* **127**(1), 75 (1987); b) Annan, T.A.; J.Gu Tian, Z.; Tuck, D.G. *J. Chem. Soc., Dalton Trans.* 3061 (1992).

142. Geloso, C.; Kumar, R.; López-Grado, J.; Tuck, D.G. *Can. J. Chem.* **65**, 928 (1987).

143. a) Castro, R.; Duran, M.L.; García-Vázquez, J.A.; Romero, J.; Sousa, A.; Castiñeiras, A.; Hiller, W.; Strahle, J. *J. Chem. Soc., Dalton Trans.* 531 (1990); b) Durán, M.L.; Romero, J.; García-Vázquez, J.A.; Castro, R.; Castiñeiras, A.; Sousa, A. *Polyhedron* **10**, 197 (1991).

144. a) Castro, R.; Durán, M.L.; García-Vázquez, J.A.; Romero, J.; Sousa, A.; Castiñeiras, A.; Hiller, W.; Strahle, J. *Z. Naturforsch. Teil B* **45**, 1632 (1990); b) Castro, R.; Durán, M.L.; García-Vázquez, J.A.; Romero, J.; Sousa, A.; Castiñeiras, A.; Hiller, W.; Strahle, J. *Z. Naturforsch. Teil B* **47**, 1067 (1990).

145. a) Castro, R.; García-Vázquez, J.A.; Romero, J.; Sousa, A.; Hiller, W.; Strahle, J. *Polyhedron* **13**(2), 273 (1994); b) Castro, R.; García-Vázquez, J.A.; Romero, J.; Sousa, A.; Prichard, R.; McAulife, C.A. *J. Chem. Soc., Dalton Trans.* 1115 (1994).

146. Shirshova, L.V.; Korableva, L.G.; Astakhova, A.S.; Lavrentiev, I.P.; Ponomarev, V.I. *Koord. Khim.* **16**(3), 348 (1990).

147. a) Sanmartín, J.; Bermejo, M.R.; García-Vázquez, J.A.; Romero, J.; Sousa, A. *Transition*

Met. Chem. **18**, 528 (1993); b) Sanmartín, J.; Bermejo, M.R.; García-Vázquez, J.A.; Romero, J.; Sousa, A. *Synth. React. Inorg. Met.-Org. Chem.* **23**(8), 1259 (1993).

148. Sanmartín, J.; Bermejo, M.R.; García-Vázquez, J.A.; Romero, J.; Sousa, A. *Transition Met. Chem.* 19, 209 (1994).

149. a) Yang, C.H.; Lin, S.F.; Chen, H.L.; Chang, C.T. *Inorg. Chem.* **19**(11), 3541 (1980); b) Yang, C.H.; Chang, C.T. *J. Chem. Soc., Dalton Trans.* 2539 (1982); c) Griffits, L.; Straughan, B.P.; Gardiner, D.J. *J. Chem. Soc., Dalton Trans.* 1193 (1983).

150. Petit, M.A.; Plichon, V.; Belkacemi, H. *New J. Chem.* **13**, 459 (1989).

151. Petit, M.A.; Thami, Th.; Even, R. *J. Chem. Soc., Chem. Commun.* 1059 (1989).

152. Petit, M.A.; Bouvet, M.; Nakache, D. *J. Chem. Soc., Chem. Commun.* 442 (1991).

153. Bastida, R.; De Blas, A.; Fenton, D.E.; Rial, C.; Rodríguez, T.; Sousa, A. *J. Chem. Soc., Dalton Trans.* 265 (1993).

154. Bastida, R.; Lage, T.; Parrado, C.; Rodríguez, T.; Sousa, A.; Fenton, D.E. *J. Chem. Soc., Dalton Trans.* 2101 (1990).

155. Bastida, R.; González, S.; Rodríguez, T.; Sousa, A.; Fenton, D.E. *J. Chem. Soc., Dalton Trans.* 3643 (1990).

156. Kharisov, B.I.; Gójon-Zorrilla, G.; Blanco, L.M.; Montoya, F.C.; Burlov, A.S.; Garnovskii, D.A.; Vasil'chenko, I.S.; Bren, V.A.; Garnovskii, A.D. *Koord. Khim.* **22**(5), 132 (1996), (*Proc. XVIII Chugaev Conference*, Moscow).

157. Kharisov, B.I.; Gójon-Zorrilla, G.; Garnovskii, A.D.; Blanco, L.M.; Dieck, T. *Rev. Soc. Quím. Mex.* **39**(5), 292, 349 (1995), (*Proc. XXXI Mexican Congress on Chemistry*).

158. Coronado, G.; Blanco, L.M.; Kharisov, B.I.; López, R.L. *Rev. Soc. Quím. Méx.* **40**(special issue), 110, 185, 186 (1996) (*Proc. XXXII Mexican Congress on Chemistry*).

159. Durán, M.; García-Vázquez, J.A.; Romero, J.; Castiñeiras, A.; Sousa, A.; Garnovskii, A.D.; Garnovskii, D.A. *Polyhedron* **16**(10), 1707 (1997).

160. Romero, J.; García-Vázquez, J.A.; Durán, M.; Castiñeiras, A.; Sousa, A.; Garnovskii, A.D.; Garnovskii, D.A. *Acta Chem. Scand.* **51**, 672 (1997).

161. Kharisov, B.I.; Blanco, L.M.; García-Luna, A. *Rev. Soc. Quím. Méx.* (1999), accepted for publication.

162. Castro, R.; García-Vázquez, J.A.; Romero, J.; Sousa, A.; Castiñeiras, A.; Hiller, W.; Strahle, J. *Inorg. Chim. Acta* **211**, 47 (1993).

163. Annan, T.A.; Kumar, R.; Tuck, D.G. *Inorg. Chem.* **29**, 2475 (1990).

164. Habeeb, J.J.; Tuck, D.G.; Zhandire, S. *Can. J. Chem.* **57**, 2196 (1979).

165. Khan, M.A.; Kumar, R.; Tuck, D.G. *Polyhedron* **7**, 49 (1988).

166. Hencher, J.L.; Khan, M.A.; Said, F.F.; Tuck, D.G. *Polyhedron* **4**, 1263 (1985).

167. Castro, R.; Durán, M.L.; García-Vázquez, J.A.; Romero, J.; Sousa, A.; Castellano, E.E.; Zukerman-Schpector, J. *J. Chem. Soc., Dalton Trans.* 2559 (1992).

168. a) Chandha, R.; Kumar, K.; Tuck, D.G. *J. Chem. Soc., Chem. Commun.* 188 (1986); b) Chandha, R.; Kumar, K.; Tuck, D.G. *Polyhedron* **7**(12), 1121 (1988).

169. Castro, R.; Durán, M.L.; García-Vázquez, J.A. et al. *Polyhedron* **11**, 1195 (1992).

170. Habeeb, J.J.; Tuck, D.G. *J. Organomet. Chem.* **146**, 213 (1978).

171. Osman, A.; Tuck, D.G. *J. Organomet. Chem.* **169**, 255 (1979).

172. Mengoli, G.; Daolio, S. *J. Chem. Soc., Chem. Commun.* 96 (1976).

173. Habeeb, J.J.; Tuck, D.G. *J. Organomet. Chem.* **134**, 363 (1977).
174. Habeeb, J.J.; Tuck, D.G. *J. Chem. Soc., Dalton Trans.* 696 (1976).
175. Habeeb, J.J.; Osman, A.; Tuck, D.G. *Inorg. Chim. Acta* **35**(1), 105 (1979).
176. Said, F.F.; Tuck, D.G. *Can. J. Chem.* **58**, 1673 (1980).
177. Habeeb, J.J.; Tuck, D.G. *J. Organomet. Chem.* **139**, C17 (1977).
178. Casey, A.T.; Vecchio, A.M. *Appl. Organomet. Chem.* **4**, 513 (1990).
179. Kumar, R.; Tuck, D.G. *J. Organomet. Chem.* **281**(2), C47 (1985).
180. Shirokii, V.L.; Sutormin, A.B.; Maier, N.A.; Ol'dekop, Yu.A. *Zh. Obshch. Khim.* **53**(8), 1892 (1983).
181. Ol'dekop Yu.A.; Maier N.A.; Shirokii V.L. et al. *Izv. Akad. Nauk BSSR, Ser. Khim.* (3) 89 (1981).
182. Ol'dekop, Yu.A.; Maier, N.A.; Shirokii, V.L. et al. *Zh. Obshch. Khim.* **52**(8), 1918 (1982).
183. Grobe, J.; Schneider, B.N.; Zimmermann, H. *Z. Anorg. Allg. Chem.* **B481**(1), 107 (1981).
184. Ohst, H.H.; Kochi, J.K. *J. Chem. Soc., Chem. Commun.* 121 (1986).
185. Ohst, H.H.; Kochi, J.K. *J. Am. Chem. Soc.* **108**(11), 2897 (1986).
186. Ohst, H.H.; Kochi, J.K. *Inorg. Chem.* **25**(12), 2066 (1986).
187. Tomilov, A.P.; Floshin, M.Ya.; Smirnov, V.A. *Electrochemical Synthesis of Organic Compounds.* Leningrad: Khimiya (1976), pp. 390–391.
188. Teodoradze, G.A. *J. Organomet. Chem.* **88**(1), 1 (1975)
189. Blanco, L.M.; Kharisov, B.I.; Garnovskii, A.D. *Proc. XXXII Int. Conf. on Coordination Chemistry*, Santiago-de-Chile (1997), p. 33.
190. Kharisov, B.I.; Blanco, L.M.; Salinas, M.V.; Garnovskii, A.D. *J. Coord. Chem.* **47**, 135 (1999).
191. Garnovskii, A.D.; Burlov, A.S.; Garnovskii, D.A. et al. *Koord. Khim.* **23**(5), 399 (1997).
192. Garnovskii, D.A.; Romero, J.; García-Vázquez, J.A.; Duran, M.L.; Castiñeiras, A.; Sousa, A.; Burlov, A.S.; Garnovskii, A.D. *Koord. Khim.* **24**(1), 1 (1998).
193. Garnovskii, D.A. et al. *Russ. Chem. Bull.* **45**(8), 1988 (1996).
194. Burlov, A.S.; Kuznetsova, L.I.; Garnovskii, D.A.; Kharisov, B.I.; Blanco, L.M.; Lukov, V.V.; Garnovskii, A.D. *J. Coord. Chem.* **47**(3), 467 (1998).
195. Rodríguez, A.; Romero, J.; García-Vázquez, J.A.; Durán, M.L.; Sousa-Pedrares, A.; Sousa, A.; Zubieta, A. *Inorg. Chim. Acta.* **284**, 133 (1999).
196. Pérez-Lourido, P.; García-Vázquez, J.A.; Romero, J.; Louro, M.S.; Sousa, A.; Zubieta, J. *Inorg. Chim. Acta.* **271**, 1 (1998).
197. Pérez-Lourido, P.; Romero, J.; García-Vázquez, J.A.; Sousa, A.; Zubieta, J.; Maresca, K. *Polyhedron.* **17**(25–6), L1457 (1998).
198. Pérez-Lourido, P.; García-Vázquez, J.A.; Romero, J.; Sousa, A. *Inorg. Chem.* **38**(3), 538 (1999).
199. Kharisov, B.I.; Blanco, L.M.; Torres-Martínez, L.M.; García-Luna, A. *Ind. Engineer. Chem. Res.* in press (1999).

Chapter 3

Oxidative Dissolution of Metals and Metal Oxides in a Liquid Phase

3.1 INTRODUCTION

Studies of the theory of solid–liquid interaction are becoming more essential in view of the necessity to increase the productivity of technological processes in the chemical, hydrometallurgical, chemopharmaceutical and other branches of industry. Broadening of the field of application of hydrometallurgy, which in some ways is preponderant over pyrometallurgy, draws attention to dissolution processes. Metals and their oxides, which are often natural products and components of minerals, are widely used in engineering and technology, in particular in powder metallurgy and metal-ceramics production. At the same time, there is no common approach nowadays to studying the kinetics of dissolution of metal powders and oxides. Therefore, before consideration of reactions in the liquid phase, it is necessary to study kinetics by the known routes.

3.2 KINETICS OF DISSOLUTION OF METAL POWDERS AND METAL OXIDES

Several methods [1–3] are employed to study solid dissolution kinetics. The most applicable are the two techniques that involve powder samples or samples of some definite shape (pressed pellets, plates, cylinders or disks) with a known surface. The so-called revolving disk method was applied successfully to study the kinetics of metal and semiconductor dissolution [1,2]. The methods based on the use of a compact sample, however, do not produce unambiguous results to characterize the kinetics of a real hydrometallurgy process. It is difficult to prepare the compact sample with a precisely known surface; besides it readily loses its initial properties, becoming covered with the interaction products or with a network of cavities, thus changing the surface area.

The methods using powder samples reveal more exactly the character of hydrometallurgy process kinetics. Difficulties in the mathematical description of these processes arise from in the complexity of the time-dependent topochemical reaction rate change. This rate is determined by the characteristics of the appearance and growth of the solid product nuclei, which change the surface area of the interphase, as well as by influence of the interaction products.

One of the most widely used methods to describe hydrometallurgy processes is that of "initial sites" employing the Shchukarev–Dolivo-Dobrovolsky equation [4–6]. It states that the reaction surface is free from reaction products and the concentration of active centers is constant in the initial period of the interaction [1,7,8].

For example, the rate of dissolution of zinc oxide in KOH solutions changes with time, being maximum in the initial reaction period [9]. Using the dependence of the rate on the KOH concentration (1–10 $mol\,dm^{-3}$), the reaction was determined to be first-order on hydroxyl ions. The dependence plotted in the Arrhenius equation mode has a form similar to a Zeldovich curve [10] and the activation energy (E_{act}) upon a rise of temperature from 293 to 323 K diminishes from 52 to 15 $kJ\,mol^{-1}$. Thus, in the temperature range studied, a change in the nature of the rate-limiting step is observed that in the authors' opinion [9] is explained by the increase in diffusion hindrance.

To ascertain the possible isolation of cobalt from admixtures of pyrrhotine concentrates and slags, the kinetics of dissolution of Co_3O_4 in hydrochloric acid has been studied [11]. It is stated that the time dependence of the dissolution rate is similar to that described in [9], the reaction order on HCl (1–5.6 $mol\,dm^{-3}$) equals 1, and E_{act} in the temperature range 303–353 K is 5 $kJ\,mol^{-1}$ and is constant.

Another example of the method of "initial sites" appears in a kinetic analysis of experimental data obtained on the interaction of Cu_2O and CuO with aqueous solutions of sodium ethylenediaminetetraacetate (EDTA) [12]. For the preliminary evaluation of the reactivity of these oxides, their thermodynamic characteristics were compared.

From the lower value of formation heat and isobaric potential of copper(I) oxide in comparison with copper(II) oxide, the reactivity of the latter was supposed to be smaller. The dissolution kinetics was studied with polydispersed powders with specific surfaces of 2.0 (Cu_2O) and 12.6 $m^2 g^{-1}$ (CuO). It was stated that reaction orders on EDTA (0.01–0.05 mol dm^{-3}) equaled 1 for both oxides. The values of the specific rates of dissolution confirmed the assumption made from the analysis of the thermodynamic characteristics and showed that the rate of dissolution of Cu_2O is 700 times greater than that of CuO. The values of the activation energies (26 kJ mol^{-1} for Cu_2O, 71 kJ mol^{-1} for CuO), calculated by means of the Arrhenius equation, indicate [12] that in the temperature range 298–338 K CuO dissolution appears in the kinetic region, and that of Cu_2O in the diffusokinetic one.

The maximum rate of dissolution in the initial period of interaction observed in the studies mentioned above may demonstrate, in the authors' opinion [10,13], that the whole surface of the particle becomes simultaneously covered with nuclei of a solid product which forms a compact layer. In this case the reaction kinetics is to be determined only by the rules of interfacial motion and may be described with the "compressing sphere" equation (3.1):

$$1 - (1 - \alpha)^{1/3} = kt \qquad (3.1)$$

where α is degree of conversion, k the dissolution rate constant and t the time. The equation may be applied for the description of σ-shaped kinetic curves with good approximation if the thickness of the compact layer formed is increased over time, and its diffusion resistance is negligible [10]. Depending upon the moment of interface formation, equation (3.1) may describe either the whole curve or part of it (after the rate maximum).

The "compressing sphere" model was used, for example, to describe the kinetics of the interaction of metal aluminum with aqueous ethanolic solutions of hydrochloric acid [14]. Experimental dependence of the degree of conversion upon time has a σ-shape and readily becomes rectilinear in $(1 - (1 - \alpha)^{1/3})$ versus t coordinates up to high (more than 0.9) degrees of conversion. Dissolution rate constants were used to determine E_{act}, which was equal to 44 kJ mol^{-1} at 298–328 K.

Satisfactory results were obtained in describing the kinetics of the interaction of uranium dioxide with solutions of potassium cyanide, sodium carbonate and bicarbonate [15] by equation (3.1). Rectilinear dependences in coordinates $(1 - (1 - \alpha)^{1/3})$ versus t allowed the authors to conclude that the reaction product did not limit the dissolution rate until $\alpha = 0.8$ and the interface moved at a constant rate. E_{act} at temperatures of 303–352 K is equal to 51.1 kJ mol^{-1}.

Influences of various factors on the dissolution rate constant of copper(II) oxide in EDTA and ammonia solutions, determined with equation (3.1), have been studied [16,17]. It was shown that the dissolution rate depended on the pH and ammonia concentration. It was also shown that on changing the EDTA concentration, the rate had a maximum. The authors connect this dependence with the adsorption of excess EDTA on the oxide surface.

The dissolution of oxides and hydroxides of iron (magnetite, maghemite, goethite, lepidocrocite and akaganeite) in hydrochloric and perchloric acids has been studied [18]. It was shown by electron microscopy that dissolution proceeded isotropically. In this case the dissolution rate is proportional to the surface area and can be described by equation (3.1), which was confirmed experimentally. The values of E_{act} calculated by the Arrhenius equation are equal to 67–94 kJ mol^{-1} (283–333 K).

Kinetic curves that were s-shaped with different induction periods were obtained by treatment of the results of the interactions of the oxides of copper, manganese, cobalt, nickel and iron oxide interactions with aqueous hydrochloric, sulfuric and oxyethylidenediphosphonic acids [19–23]. It was shown that the "compressing sphere" equation described kinetic curves satisfactorily at $\alpha > 0.2$–0.6. The authors [19,23] explained such a delay of dissolution by the lack of active centers in the initial period, and by the reaction front spreading onto the whole surface by a definite moment in time.

When a reacting surface develops irregularly, the dissolution rate is determined by two factors [13]: the rate constant of the active centers germinating (k_g) and the rate constant of their dissolution (k_i), calculated from the equations

$$k_i = r_0 \frac{t^*_{0.5}}{t_{0.5} - t_e}$$

$$k_g = \frac{Ak_i}{4\pi r_0^3}$$

where r_0 is a spherical radius of a particle, t_e is the induction period from the equation $\alpha^{1/4} = \text{constant} \times (t - t_e)$. The coefficient A is found by plotting the experimental data in α versus $(t - t_e)/(t_{0.5} - t_e)$ coordinates onto the net of theoretical curves (in the same coordinate system) with different A values. The value of $t_{0.5}$ for $\alpha = 0.5$ is found using the A value [13].

This model was used to investigate the kinetics of dissolving CuO, Fe_2O_3 and Fe_3O_4 in sulfuric acid [24,25]. It was found that both dissolving and active-center germinating rates rose as the acid concentration increased. In the case of iron oxides, the dissolving rate of magnetite was approximately one order more than that of hematite, while the active-center germinating rates were approximately equal. When the content of iron(II) ions in solution increased, both dissolving and active-center germinating rates rose. The interaction kinetics in this case was described by equation (3.1). The coefficient A changed from 30 for copper(II) oxide to 1×10^3–5×10^4 for Fe_3O_4.

The equations used are applicable to processes with kinetics determined mainly by surface phenomena. Anomalous rapid dissolution of oxides in acids is explained by the predominance of volume processes, when the increase in active centers in a volume of solid oxide is caused by diffusion of hydrogen ions [26]. In this case thread-like branching of active-center germs develops in the direction perpendicular to the sample surface [13]. Under such irregular development of the reaction front into the volume, the change in the dissolved oxide moiety is described by equation (3.2) [13], where A is a kinetic constant, sh is hyperbolic sinus, $t^* = k_i t$.

$$-\ln(1 - \alpha) = A sh\, k_i t = A sh\, t^* \qquad (3.2)$$

The applicability of equation (3.2) and the dominance of volume processes (up to $\alpha = 0.6$) were shown in kinetic studies of some metal oxides dissolved in sulfuric acid [21,22]. In finding an analytical type of equation, the authors proceeded from the assumption that the rate of the oxide phase conversion is determined by germinating and dissolving active centers in the oxide volume. Values of A in equation (3.2) were found to be independent of the acid concentration and are determined only by the nature of the oxide.

Equation (3.2) was used to treat results of a calorimetric study of the kinetics of lead(II) oxide interaction with aqueous nitric acid [27]. An E_{act} value of 31.7 kJ mol^{-1} was determined from the temperature dependence of the rate constant (303.5–334.9 K).

As mentioned above, the "compressing sphere" equation satisfactorily describes kinetic curves, if on dissolution the layer of reaction products does not hinder diffusion. On formation of a compact nonporous layer of the products, the rate of diffusion is less than that of chemical reaction. In this case equation (3.3), suggested by Ginstling and Brownstein [28], is applicable.

$$1 - (\tfrac{2}{3})\alpha - (1 - \alpha)^{2/3} = kt \qquad (3.3)$$

The interaction of MnO_2 with a dimethyl sulfoxide solution of SO_2 [29] proceeds with the formation of a layer of intermediate products on the surface, and equation (3.1) becomes inapplicable. The kinetic curves obtained become rectilinear in the $(1 - (2/3)\alpha - (1 - \alpha)^{2/3})$ versus t system of coordinates, confirming the conclusion about the decisive influence of inner diffusion on the interaction studied.

On dissolving zinc sulfide in an aqueous solution of iron(III) chloride, it was found that a layer of free sulfur was formed on the surface of the reacting particle, thus making equation (3.3) applicable [30]. The given equation satisfactorily describes the process of dissolution complicated by the diffusion through the interaction product layer. According to [7], the value of $E_{act} = 90.0 \pm 12.5\,kJ\,mol^{-1}$ (303–343 K) indicates the kinetic nature of the rate-limiting step.

The conclusion about the presence of diffusion inhibition in the solid (DIS) appearing after the rate maximum (for s-shaped curves) may be made by comparing the rate constants before and after the maximum rate [10]. Supposing independent growth and simultaneous contacts of solid product nuclei, as well as full packing of the particle surface with hemispherical nuclei, the rate constants in different parts of the kinetic curves may be calculated with a quite simple formula using only one parameter of the rate maximum [10]. When DIS is absent, the two rate constants must be equal. This approach was used in the study of the the interactions of solid and liquid reagents, forming insoluble reaction products [10]. For example, in a study of the interaction kinetics of iron metal and water, it was shown that diffusion processes in the solid do not interfere even after the formation of an iron(II) hydroxide layer on the surface of the reacting particle. The rate constants before and after the rate maximum were the same, and E_{act} (125 $kJ\,mol^{-1}$) was constant over the whole range of degrees of conversion. In contrast, the forming hydroxide layer produced a notable diffusion hindrance in the interaction of aluminum powder with water in the presence of NaOH [31], since the rate constants after the maximum were two to three times less than before it. The authors supposed that in this case NaOH reacted only with hydroxide, thus assisting the main reaction.

In autocatalytic reactions, contrary to the previous examples, forming products do not hinder, but rather accelerate, the interaction process [7,32,33]. When the reaction has an autocatalytic mechanism, the dissolution rate is proportional to both the quantity of the product formed, and the quantity of undissolved solid phase. The process is described by equation (3.4), where B is a kinetic constant.

$$\ln[\alpha/(1 - \alpha)] = kt + B \qquad (3.4)$$

Thus, investigating titanium dioxide and metal titanates dissolved in sulfuric acid, s-shaped kinetic curves were obtained [32]. The rise of dissolution rate in the initial period is determined by the formation of TiO^+ ions. This is confirmed by the noticeable rate increase in the presence of strong reducing agents. For example, the reaction rate is one order greater in the presence of zinc powder. The kinetic curves of titanium dioxide dissolution are satisfactorily rectilinear in $\ln[\alpha/(1 - \alpha)]$ versus t coordinates. Using equation (3.4), the authors calculated kinetic constants k and B, and ascertained the factors influencing their values. The rate constant was found to be independent of hydrogen ion concentration, and the reaction order was zero. The acid concentration affected only the B value, which characterizes the surface state.

The influence of pH on Fe_3O_4 interaction with EDTA solutions and oxyethylidenediphosphonic acid was studied [33]. The s-shaped kinetic curves as well as complex stability constants influencing the reaction rate enabled the authors to suggest an autocatalytic mechanism for the process and to use equation (3.4) for its description.

An important step in the theory of topochemical kinetics has been made with the use of general kinetic equation (3.5) [10,34],

$$\alpha = 1 - \exp(-Kt^n) \qquad (3.5)$$

where n and K are kinetic parameters. Equation (3.5) can be regarded as a probability function of the degree of conversion in processes of general character. It was applied to analyze the change in quantity and composition of biological and microbiological populations, processes of tumor growth, auto-oscillation and other processes in biosystems [34]. The descriptive potential of equation (3.5) is wide, and its applications ought to be sensed with its physical meaning [10].

This equation has some advantages over the ones mentioned above, for it describes the process before and after the rate maximum with the same values of n and K. There are, however, some systems where n and K are changing during the reaction and with a change in the reaction conditions. Thus, when copper(II) and iron(III) oxides are dissolved in sulfuric acid [35,36], the values of n and K change during the reaction. In this case, the s-shaped kinetic curves become rectilinear in coordinates $\log[-\log(1-\alpha)]$ versus $\log t$, and the dependences obtained have some linear regions: two for CuO and three for Fe_2O_3. Different regions of the kinetic curves should be described by the function having different parameters, which may be caused by a change in the reaction mechanism during the process. This may lead to some difficulties in analyzing the results obtained.

As a rule, it is difficult to determine the influence of reaction conditions on each of the kinetic parameters because both can be changed under the action of the investigated factors, such as initial dispersion of the sample, how it was obtained and its storage. The mutual variation of kinetic parameters of equation (3.5) may be explained by compensation dependence [37,38].

The physical sense of the coefficient n was interpreted by Erofeev for thermal decomposition reactions as the sum of sequencing steps in the formation of stable active centers, and a constant which characterizes the shape of the developing nuclei of reaction products [39,40]. However, the experimental n value may appear to be nonintegral, due to the different nature of the active centers on which nuclei of a solid grow. In accordance with [41], the n value gives the most general view of the reaction mechanism. Thus, high n values are characteristic for the case of rapid generation of new active centers and rapid nonlinear growth of solid nuclei. It was shown [41] that equation (3.5) can be applied for the description of processes in the diffusion region when $n < 1$. Since the diffusion may influence the rate of the process in different ways, it was suggested that the value of $(1-n)$ should be considered as a measure of entry into the diffusion region.

Thus, the formation of solid in the reaction is the precondition of the appearance of DIS, which may change during the reaction and depend on experimental conditions. The main indications of diffusion inhibition are lowering of the rate constant after the maximum and low values of n in the general topochemical equation. In [42–67] the authors attempted to elucidate the influence of reaction product properties on the change in diffusion inhibition and to correlate the influencing factors. The kinetics of interaction of copper, nickel, cobalt, zinc, cadmium, and lead powders and their oxides with aqueous and nonaqueous (alcohols, acetonitrile, dimethylformamide, dimethyl sulfoxide) solutions of ammonium salts (chloride, bromide, iodide, thiocyanate, nitrate, acetate) has been investigated. Electron-microscopic investigations of metal samples treated with solutions of ammonium salts in different conditions showed that the interaction began at definite points of metal surface (active centers), where the solid product nuclei of nearly spherical form grow [50,59]. Their growth results in the formation of a compact layer on the particle surface, which in some cases can be observed microscopically. Formation of a slightly soluble product layer on the metal surface, sometimes of sizeable thickness, was confirmed by an Auger spectral study [59,61]. The composition of the supposed interaction products could be found in some cases. Thus, the compound formed by the treatment of lead powder pellets with aqueous ammonium chloride solution contains chloride and oxygen ions. On etching the layer with argon ions, chloride and oxygen were removed simultaneously. Furthermore, the content of chlorine and oxygen, calculated from Auger spectra, remained approximately the same during the etching. It was concluded [59] that oxy- or hydroxychloride of lead formed on the surface of the initial sample had a definite stoichiometric composition with $ClO = 1:1$. The formation of CuX ($X = Cl, Br, I, SCN$) compounds on the surface of the reacting particle was found for copper [61]. The product compact layer formation can hinder the passage of the starting reagents and removal of the reaction products, and cause DIS. Evaluation of the product density on the surface using the Pilling–Bedwords criterion [7] has shown that the ratio of molar volumes of the reaction products and the metal or its oxide is always more than 1. This means that the packing of the layer is sufficiently compact during its formation. The layer density on copper or copper(I) oxide increases in the sequence $Cl < Br < I < SCN$.

Analysis of the experimental data confirms the assumption about the important role of DIS in the kinetic behavior of metal powders and their oxides. The degree of influence of DIS depends on the character of the system and external factors, and may change during the reaction run. Changes in external factors can substantially transform the shape of the kinetic curves; for example, a σ-shaped curve changes into an s-shaped one at the rate maximum. The position of the maximum is influenced by the temperature or the ammonium salt concentration [54,58,60,61]. Such influences may be explained by DIS and the form of the growing nuclei [10]. An increase in diffusion inhibition may lead to distortion of the spherical shape of the nuclei, which in extreme cases can localize in a solid monolayer on the surface.

The reaction rate increases with temperature far more than the diffusion coefficient, so generally the temperature rise accelerates the formation of the solid and strengthens DIS, diminishing α_{max}. For example, the rate maximum completely disappears in the system of Cu–NH$_4$SCN–AN [61] for a temperature rise from 293 to 323 K. There are systems (PbO–NH$_4$SCN–DMF, PbO–NH$_4$NO$_3$–DMSO, or alcoholic solutions of NH$_4$Ac) where the rise in temperature leads to increasing α_{max} [54,58]. This may be explained by an increase in the solubility of the products with the rise in temperature which may weaken the DIS influence. In the interaction of Cu$_2$O with ammonium salt solutions in DMF, α_{max} diminishes in all cases, and sometimes disappears completely with the temperature rise and concentration increase [66]. Thus, the increase in concentration, similarly to the rise in temperature, promotes accumulation of reaction products on the surface. On the other hand, the increase in ammonium salt concentration may give soluble anionic complexes and lead to thinning of the surface layer. In the interaction of copper powder and methanolic solutions of ammonium thiocyanate of low concentration, when DIS is considerable, the rate drops to zero at $\alpha = 0.5$. Every particle of the copper becomes covered with a compact layer of copper(I) thiocyanate, hindering diffusion of the reagents into the reaction zone. On increasing the ammonium thiocyanate concentration the kinetic curve becomes s-shaped with the rate maximum [61].

After the rate maximum, the diminishing reacting surface leads to a decrease in the rate. In the same part of the kinetic curve, the properties of the product layer may change (e.g., density or thickness). This can lead to a rate constant decrease after the maximum. Thus, it is assumed that for the systems with s-shaped kinetic curves the rate constant before the maximum depends on the character of the system, and the nature of the solvent in particular. The rate constant after the maximum is determined by the properties of the interaction product layer. Calculations of rate constants in different parts of kinetic curves were accomplished using the approach described in [10]. As Figure 3.1 shows, the systems studied included some with linear dependence of rate constant on the solvent donor number.

The criterion of DIS presence after the maximum is the factor of diffusion inhibition f (the ratio of rate constant k_2 after the maximum, to k_1 before it) [10]. From the rate constants in Table 3.1 [10], f values are about 1 for good solvating solvents. This indicates a weak influence of DIS after the maximum. The rate constant after the maximum decreases (up to 30%) for solutions in alcohols, of lower solubility. The solubility of products may change

TABLE 3.1
Rate constants (mol $m^{-2} min^{-1}$) of metal powders and metal oxides dissolving in different parts of the kinetic curves [59,65,66]

T, K	C_{NH_4X}, mol dm^{-3}	k_1	k_2	f
Pb–NH₄Ac–DMF				
295	0.10	0.097	0.099	1.02
295	0.12	0.116	0.118	1.02
303	0.10	0.119	0.116	0.97
PbO–NH₄I–DMSO				
303	0.50	0.048	0.050	1.04
298	0.50	0.034	0.036	1.06
293	0.40	0.014	0.014	1.00
293	0.45	0.017	0.018	1.06
Zn–NH₄Br–DMSO				
303	0.20	0.014	0.013	0.93
313	0.20	0.012	0.011	0.92
303	0.30	0.012	0.011	0.92
293	0.40	0.009	0.009	1.00
303	0.40	0.012	0.011	0.92
Cu–NH₄SCN–CH₃OH				
293	0.80	0.018	0.016	0.89
303	0.80	0.027	0.022	0.81
323	0.80	0.062	0.049	0.79
323	1.00	0.078	0.056	0.72
PbO–NH₄Ac–C₂H₅OH				
283	0.20	0.048	0.042	0.88
283	0.15	0.051	0.038	0.75
288	0.15	0.061	0.047	0.76
277	0.15	0.031	0.026	0.84
293	0.25	0.083	0.070	0.84
PbO–NH₄Ac–i-C₃H₇OH				
284	0.15	0.041	0.034	0.83
284	0.10	0.027	0.024	0.89
284	0.13	0.037	0.029	0.78
288	0.10	0.033	0.029	0.88
292	0.10	0.046	0.039	0.85

the degree of DIS action. The solubility of CuX compounds can be characterized by their solubility product, which diminishes in the sequence Cl < Br < I < SCN to cause an increase in the degree of DIS and a decrease in the rate constants after the maximum (Fig. 3.2).

Analysis of the influence of the product layer density using the Pilling–Bedwords criterion (K) showed that there are systems with linear dependence of rate constant on K after the rate maximum (Fig. 3.2).

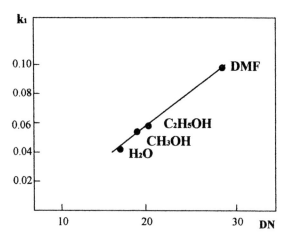

Fig. 3.1. Dependence of the velocity constant (up to the maximum velocity) on the donor number (DN) of the solvent for the Pb–NH$_4$Ac–Solv system.

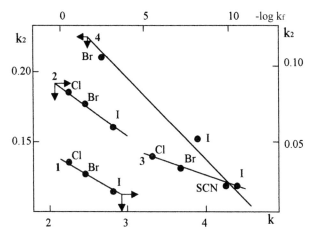

Fig. 3.2. Influence of the density (1, 2, 4) and solubility (3) of the reaction product on the velocity constant of the reaction (after the velocity maximum) in the systems Cu$_2$O–NH$_4$X–H$_2$O (1, 3), Cu$_2$O–NH$_4$X–DMSO (2) and Pb–NH$_4$X–DMSO (4). Density here is the density of the product layer evaluated using Pilling–Bedwords criterion (K) and solubility is the Solubility Product (SP) of the product of the reaction. The curves 1, 2, 4 show the dependence of K on the velocity constant [rate constant] of the reaction (after velocity [rate] maximum) and curve 3 is the dependence of log SP on the velocity consant [rate constant] of the reaction (after velocity [rate] maximum).

Meanwhile, the approach described does not allow a conclusion to be reached about the degree of DIS for kinetic curves without a rate maximum. Using the Sakovich supposition [41] about the correlation of the coefficient n with the degree of diffusion inhibition, kinetic experiments were analyzed with the help of general topochemical equation (3.5) [45,46,48,50,52–63,65–67]. This is the preferred model for analyzing the dependence of the invariance of the kinetic curves on external factors and the progress of the reaction (Fig. 3.3). The systems investigated can be conventionally divided into fully and partly invariant. For example, the kinetic curves are fully invariant with temperature and concentration in Cu_2O–NH_4X–Solv (excluding DMF) (Fig. 3.3a), CoO–NH_4SCN–Solv (excluding acetonitrile), Ni–NH_4SCN–CH_3OH, Zn–NH_4SCN-Solv and ZnO–NH_4X–Solv (excluding aqueous and methanolic solutions) systems. Infringement of invariancy during the change in temperature was marked for kinetic curves of interaction in: Cd–NH_4SCN–Solv, CdO–NH_4SCN–Solv, PbO–NH_4Br (NH_4SCN)–DMSO, PbO–NH_4Ac–CH_3OH (C_2H_5OH) and PbO–NH_4Br–H_2O systems.

The infringement of invariance to a temperature change was marked for kinetic curves describing the interaction in the Cu–NH_4NO_3–DMF, Cu_2O–NH_4Br–DMF, Zn–NH_4Br–DMSO (Fig. 3.3b), ZnO–NH_4I (NH_4SCN)–CH_3OH, CdO–NH_4SCN–C_2H_5OH, (n-C_3H_7OH), PbO–NH_4NO_3–DMF (DMSO, H_2O), Pb–NH_4NO_3–DMSO and PbO–NH_4SCN (NH_4Ac)–DMF systems. The result of the study of Cu_2O interactions with dimethylformamide solutions of ammonium salts [66] shows that upon a rise in temperature the value of n diminishes (for example, for NH_4Br from 1.6 to 1.2). This may be explained by strengthening of DIS and by the transition of the process into the region of inner-diffusion. This conclusion is proved by calculations of E_{act} by means of the Arrhenius equation. Dependences plotted in $\log k$ versus $1/T$ coordinates are similar to Zeldovich curves (see Fig. 3.4). With the increase in temperature the process moves from the intermediate region into the inner-diffusion one for thiocyanate, while for ammonium bromide and iodide it moves from the kinetic region into the intermediate one between the kinetic and inner-diffusion regions.

Infringement of invariance with a change of ammonium salt concentration was marked for kinetic curves of the following systems: Cu–NH_4SCN–CH_3OH (Fig. 3.3c), Cu_2O–NH_4Br–DMF, ZnO–NH_4Br–CH_3OH, PbO–NH_4I–DMF, PbO–NH_4Cl (NH_4SCN)–H_2O, PbO–NH_4Ac–n-C_3H_7OH, Pb–NH_4NO_3 (NH_4I)–DMF, Pb–NH_4Br (NH_4Ac)–DMSO, Pb–NH_4Cl (NH_4NO_3, NH_4SCN)–H_2O. An increase in concentration, as a rule, leads

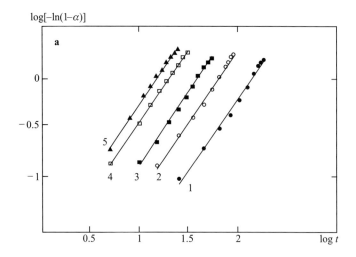

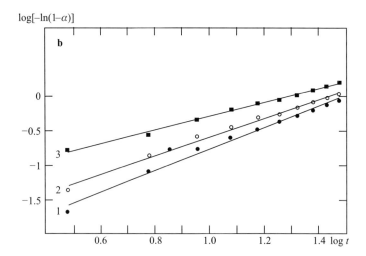

Fig. 3.3. Kinetic curves of dissolution of: (a) Cu_2O in aqueous solution of NH_4Cl (1.50 mol dm^{-3}) at 293 K (1), 303 K (2), 313 K (3), 323 K (4), 333 K (5); (b) zinc powder in the solution of NH_4Br in DMSO (0.20 mol dm^{-3} at 293 K (1), 303 K (2), 313 K (3); (c) copper powder in the solution of NH_4SCN in CH_3OH (323 K) at C (mol dm^{-3}) = 0.30 (1), 0.40 (2), 0.50 (3), 0.60 (4), 0.80 (5), 1.00 (6); (d) ZnO in aqueous solutions of NH_4Br at 323 K (1, 3), 333 K (2, 4) and C (mol dm^{-3} = 0.30 (1, 2), 0.50 (3, 4).

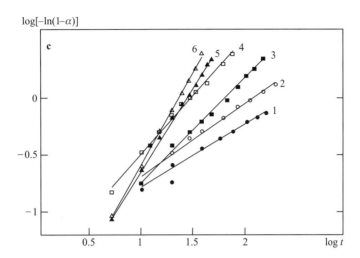

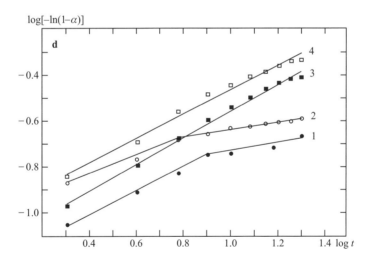

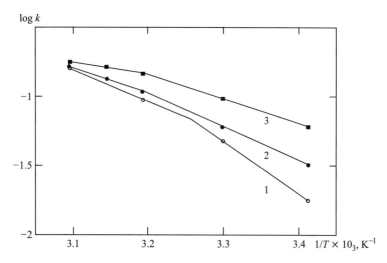

Fig. 3.4. Dependence of the velocity constant of dissolution of Cu_2O in DMF solutions of ammonium iodide (1), bromide (2) and thiocyanide (3) on temperature. C $(mol\,dm^{-3} = 0.30$ (1), 0.20 (2, 3).

to an increase in n, which may be caused by lessening of the product layer at the expense of formation of more soluble anionic complexes. For example, a concentration rise from 0.3 to 1.0 $mol\,dm^{-3}$ in the Cu-NH_4SCN-CH_3OH system leads to an increase in the n value from 0.6 to 1.6 (Fig. 3.3b). In this case, weakening of DIS upon an increase in concentration is also confirmed by the appearance of the rate maximum mentioned above.

Infringement of invariance in some systems with lead and its oxide was observed during changes of both temperature and concentration [60].

Processes in the following systems exhibit infringement of invariance during the course of the reaction: Co-NH_4Cl (NH_4Br)-Solv (excluding ethanol), Ni-NH_4SCN-Solv, Zn-NH_4Cl (NH_4Br)-H_2O, ZnO-NH_4X (X = Cl, Br, I)-H_2O (Fig. 3.3d), PbO-NH_4NO_3-DMF, PbO-NH_4NO_3 (NH_4SCN)-H_2O. In these cases with an α value of less than 0.5, a drop in the rate was observed, and the dependences in $log[-\ln(1-\alpha)]$ versus $\log t$ coordinates had a point of inflection. The n value in the second section (after the point of inflection) lessened, and in some systems was two to seven times less than in the first section [57,62]. Some experiments showed that diffusion inhibition in Ni-NH_4SCN-AN (C_2H_5OH), Co-NH_4Br-CH_3OH, Co-NH_4SCN-AN and PbO-NH_4NO_3-DMF systems was caused by the accumulation of the intermediates, and in Co-NH_4SCN-C_2H_5OH, PbO-NH_4Cl (NH_4Br, NH_4SCN)-

H_2O, $Ni-NH_4SCN-n-C_3H_7OH$ and $Co-NH_4Br-DMF$ (DMSO) systems by that of final interaction products on the surface [58,67]. The position of the point of inflection is influenced by changes in temperature and ammonium salt concentration. Thus, upon an increase in concentration, for kinetic curves of the processes in $Co-NH_4Br-DMF$ (DMSO) systems the α value corresponding to the point of inflection increases, and for the $ZnO-NH_4Br-H_2O$ system the point of inflection disappears completely at sufficiently high ammonium salt concentrations (0.5 $mol\,dm^{-3}$) (Fig. 3.3c), which indicates weakening of DIS due to formation of more soluble complexes.

The correlation of n values and kinetic curve shapes was mentioned during the discussion of the factors influencing the kinetic parameters. Thus, when the n value is more than 1.3–1.4, s-shaped kinetics curves become σ-shaped. As Fig. 3.5 shows, there are systems with σ-shaped curves with a linear dependence of n on K or log SP. The increase in density and decrease of solubility of products on the surface lead to DIS strengthening and diminution of the n value in the general topochemical equation. The rate constant after the maximum reduces also.

The agreement of the kinetic characteristics obtained through different models may be demonstrated using systems with copper and copper(I) oxide (Table 3.2). Taking into consideration the E_{act} value and other kinetic features, processes may be conventionally assigned to an inner-diffusion region, including an intermediate region (intermediate between the inner-

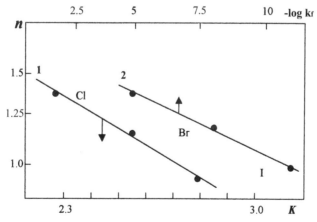

Fig. 3.5. Dependence of the coefficient n (in the generalized topochemical equation) on density (1) and solubility (2) of the reaction product for the $Cu_2O-NH_4X-H_2O$ system.

TABLE 3.2
Assignment of processes in the studied systems to kinetic regions

System	α_{max}	Influence on α_{max}		f	Invariance		n	$E_{act},^{*}$ kJ mol^{-1}	$E_{act},^{\dagger}$ kJ mol^{-1}
		T	C_{NH_4X}		T	C_{NH_4X}			
Inner-diffusion and intermediate region									
Cu$_2$O–NH$_4$SCN–H$_2$O	−			−	+	+	0.8–0.9	16	14
Cu$_2$O–NH$_4$SCN–DMSO	−			−	+	+	1.0–1.1	24	23
Cu$_2$O–NH$_4$Br–DMF	−			−	+	+	1.4–1.5	34	38
Cu$_2$O–NH$_4$Cl–DMSO	−			−	+	+	1.1–1.2	37	37
Cu$_2$O–NH$_4$Ac–DMSO	−			−	+	+	1.0–1.1	34	34
Cu$_2$O–NH$_4$SCN–AN	+	+	−	0.80–0.92	+	+	1.3–1.5	24	25
Cu$_2$O–NH$_4$NO$_3$–CH$_3$OH	+	−	−	0.94–1.09	+	+	1.8–1.9	36	36
Cu$_2$O–NH$_4$SCN–CH$_3$OH	+	−	+	0.72–0.89	+	−	0.5–1.6	29	31
Cu$_2$O–NH$_4$Ac–CH$_3$OH	+	−	−	0.95–1.08	+	+	1.6–1.8	34	35
Cu$_2$O–NH$_4$Ac–H$_2$O	+	−	−	1.00–1.10	+	+	1.6–1.7	34	34
Cu$_2$O–NH$_4$I–H$_2$O	−		−	−	+	+	0.7–0.9	34	36
Kinetic region									
Cu$_2$O–NH$_4$Cl–DMSO	+	−	−	0.97–1.00	+		1.4–1.5	48	48
Cu$_2$O–NH$_4$Br–DMSO	+	−	−	1.00–1.06	+	+	1.4–1.5	67	67
Cu$_2$O–NH$_4$I–DMSO	+	−	−	0.86–1.05	+	+	1.4–1.5	67	69
Cu$_2$O–NH$_4$SCN–DMSO	+	−	+	0.96–1.00	+	+	1.2–1.3	46	46
Cu$_2$O–NH$_4$Ac–CH$_3$OH	+	−	−	1.00–1.06	+	+	1.5–1.6	64	60
Cu$_2$O–NH$_4$Cl–H$_2$O	+	−	−	0.93–0.95	+	+	1.4	50	50
Cu$_2$O–NH$_4$Br–H$_2$O	+	−	−	0.87–1.00	+	+	1.2–1.5	44	41
Cu$_2$O–NH$_4$Ac–H$_2$O	+	−	−	0.88–1.00	+	+	1.2–1.3	52	52

TABLE 3.2 (Contd.)

System	α_{max}	Influence on α_{max}		f	Invariance		n	$E_{act},*$ kJ mol^{-1}	$E_{act},\dagger$ kJ mol^{-1}
		T	C_{NH_4X}		T	C_{NH_4X}			
Transitional region									
Cu–NH$_4$NO$_3$–DMF	−			—	−	+	1.1–1.4	130; 31	112; 44
Cu–NH$_4$SCN–DMF	−			—	+	+	1.1–1.2	22; 2	28; 3
Cu–NH$_4$NO$_3$–DMSO	−			—	+	+	1.0–1.1	109; 12	112; 10
Cu–NH$_4$SCN–H$_2$O	−			—	+	+	1.0–1.1	14; 38	—; 41
Cu$_2$O–NH$_4$Br–DMF	+	+	+	0.73–0.91	−	−	1.2–1.6	54; 41	47; 35
Cu$_2$O–NH$_4$I–DMF	+	+	+	0.75–0.90	−	−	1.2–1.5	69; 49	73; 46
Cu$_2$O–NH$_4$SCN–DMF	+	+	+	0.75–0.91	−	−	1.1–1.4	38; 13	34; 16|

*Calculated for the same degree of conversion and different temperatures.
†Calculated by the Arrhenius equation.
f, factor of diffusion inhibition, equal to k_2/k_1 (k_2 is the rate constant after the rate maximum and k_1 is the rate constant before the rate maximum).

diffusion and kinetic ones), and a kinetic region. A separate group consists of the systems of a transitional nature, where a change in E_{act} is observed in the temperature interval studied, and the transition from one kinetic region to the other usually takes place. The most satisfactory agreement of the kinetic characteristics from different models is observed for the kinetic region (E_{act} = 41–69 kJ mol^{-1}). The influence of DIS in the region is negligible, which is proved by the constant position of the rate maximum uninfluenced by changes in temperature and ammonium salt concentration, by the almost equal rate constants before and after the maximum ($f \approx 1$), and by a sufficiently high n value. Only two systems (Cu$_2$O–NH$_4$SCN–H$_2$O and Cu–NH$_4$SCN– DMSO) may be assigned to the inner-diffusion region (E_{act} = 16–24 kJ mol^{-1}), having no maximum on the kinetic curves and low n values (0.8–1.1).

Agreement of kinetic characteristics from different models is a little worse in the intermediate and transitional regions. The systems having E_{act} and thus assigned to the intermediate region can have no rate maximum. If the kinetic curve is s-shaped, the position of the rate maximum, as a rule, does not depend on temperature and concentration, the rate constants before and after the maximum are approximately equal, and n values are the maximum ones. The processes in the transition and intermediate regions may have both σ- and s-shaped kinetic curves. Contrary to the intermediate and kinetic regions, the position of the rate maximum does change with temperature and concentration for the systems of the transitional region. It is appropriate to have infringement of the invariance of the kinetics curves in the transition region. When the temperature rises, the reaction rate increases more than the diffusion rate, and E_{act} decreases to the values of the intermediate (31–41 kJ mol^{-1}) or inner-diffusion (10–16 kJ mol^{-1}) regions.

As Table 3.2 shows, only copper(I) oxide reactions are assigned to the kinetic region, while metallic copper reactions are generally in the inner-diffusion and intermediate regions. The oxidation of copper proceeds via an electrochemical mechanism [64]. The rate of this step is high; thus the influence of DIS becomes substantial. In systems with copper(I) oxide, the oxidation of copper(I) to copper(II) does not influence the general rate of the reaction. The process rate is limited by the step of chemical interaction, whose rate is smaller than DIS, so its influence is negligible.

Thus, the mode of formation, properties and behavior of the product layer can have a considerable influence on the reaction kinetics and mechanism. This is doubtless because complexation processes, whose role in the formation of the interface is so far unknown in detail, are of great importance here.

3.3 DIRECT SYNTHESIS USING HALOGENATED HYDROCARBONS

The first report on direct synthesis of coordination compounds from metal powders and nonaqueous solutions of halogenated hydrocarbons appeared in 1974 [68]. In [69–105] the authors studied the influence of the metal nature, halogenated hydrocarbons, nonaqueous solvent and an additional ligand on the character of the interaction, composition and properties of the complexes formed. The most comprehensive study was carried out for the interaction of copper and carbon tetrachloride. The principal types of compounds formed are listed in Table 3.3. The oxidation of copper in the CCl_4–DMF system gave different compounds, depending on the ratio of the initial reagents and the reaction conditions. When the volume of the oxidation mixture is sufficient, and the quantities of CCl_4 and DMF are approximately equal, a green compound, $Cu_2Cl_4(DMF)_4$, was formed. X-ray structural analysis shows that the coordination polyhedron of the copper is a strongly distorted trigonal bipyramid, with nonequivalent molecules of DMF in *cis* positions [78]. On heating to 65–70°C, the compound loses part of the DMF, being converted into yellow $CuCl_2(DMF)$ [80]. When copper powder was dissolved in a CCl_4-DMF mixture with a five-fold excess of CCl_4 by volume, yellow–brown crystals, $Cu_4OCl_6(DMF)_4$, were obtained. The monoclinic cell of the complex contains four crystalographically independent copper atoms, located at the vertices of the distorted tetrahedron. The bridging oxygen is located inside the tetrahedron and appears to be a common vertex of four trigonal bipyramids of copper coordination polyhedra [81]. When copper was oxidized with a small volume of CCl_4–DMF mixture (1:1 by volume), a dark-brown air-stable $Cu_3Cl_6(DMF)_2$ compound was formed.

Substitution of DMF by formamide (FA) or *N*-methylformamide (MFA) results in the cleavage of the C–N bond, forming ammonia and formic acid derivatives [101]. This allows one to obtain compounds containing molecules of the solvent and products of its transformation. The product of the reaction in the MFA–CCl_4 system was $[CH_3NH_3][Cu(HCOO)_3]$, and complexes $Cu(HCOO)_2(FA)_2$, $Cu_3Cl(HCOO)_5(FA)_6$, and $Cu_2Cl(HCOO)_3 \cdot 2H_2O$ were formed in FA–CCl_4. The oxidation kinetics of the processes in DMF at 287–353 K was studied with electrodialysis [83]. It was shown that the temperature dependence of the rate constant had a point of inflection. The authors supposed that the reaction rate increased considerably while the diffusion rate changed only a little. Thus, the metal surfaces become covered

TABLE 3.3
The compounds formed in interaction of copper metal with nonaqueous solutions of halogenated hydrocarbons

System	Compound	Reference
DMF–CCl$_4$	Cu$_2$Cl$_4$(DMF)$_4$, CuCl$_2$(DMF) Cu$_2$Cl$_4$(DMF)$_4$, CuCl$_2$(DMF), Cu$_4$OCl$_6$(DMF)$_4$, Cu$_3$Cl$_6$(DMF)$_2$ Cu$_4$OCl$_6$(DMF)$_4$, Cu$_3$Cl$_6$(DMF)$_2$	78, 80
MFA–CCl$_4$	[CH$_3$NH$_3$][Cu(HCOO)$_3$]	101
FA–CCl$_4$	Cu$_2$Cl(HCOO)$_3$·2H$_2$O	101
DMA–CCl$_4$	Cu$_2$Cl$_3$(OH)(DMA)$_2$	71
DMSO–CCl$_4$	CuCl$_2$(DMSO)$_2$	78
DMSO–C$_4$H$_9$Br	(CuBr$_2$)$_2$(DMSO)$_3$·H$_2$O CuBr$_2$(DMSO)$_3$·H$_2$O	71
PyO–CCl$_4$	Cu(PyO)$_2$Cl$_2$, Cu(Py)$_2$Cl$_2$	82
PyO–CHBr$_3$	Cu(PyO)$_2$Br$_2$, Cu(Py)$_2$Br$_2$, Cu(PyO)$_4$Br$_2$	82
PyO–C$_2$H$_4$Br$_2$	Cu$_4$OBr$_6$(Py)$_4$	82
TPPO–CCl$_4$–methyl ethyl ketone	Cu(TPPO)$_2$Br$_2$, Cu$_4$OBr$_6$(TPPO)$_4$	82
DMF–CCl$_4$–adenine	CuCl$_2$(A)(DMF)	98
DMF–CCl$_4$–1,2,4-triazole	CuCl$_2$(T), CuCl$_2$(DMF)(T)$_2$	98
DMF–CCl$_4$–benzotriazole	CuCl$_2$(BTA)(DMF)·2H$_2$O	98
DMF–CCl$_4$–N,N'-bis-(α-pyridyl)oxamide	Cu$_2$Cl$_4$(Pyoxam)$_3$	98
DMSO–CCl$_4$–H$_2$D	[CuCl$_2$(H$_2$D)]$_2$·DMSO	72
DMF–CCl$_4$–H$_2$D	CuCl$_2$(H$_2$D)(DMF)	72
DMF–CCl$_4$–2,2'-bipy	Cu(2,2'-bipy)Cl$_2$	72
DMSO–CCl$_4$–2,2'-bipy	Cu(2,2'-bipy)Cl$_2$	72
C$_2$H$_5$OH–CCl$_4$–H$_2$D	CuCl$_2$(H$_2$D)	72
DMF–CCl$_4$–1,5-COD	[Cu(1,5-COD)Cl]$_2$	72
DMSO–CI$_4$–TU	Cu$_3$I$_2$(Dtc)$_5$, Cu$_2$I$_4$(Dtc)$_3$	99
DMF–CI$_4$–TU	Cu$_2$I(Dtc)$_2$	99

PyO, pyridine oxide; TPPO, triphenylphosphine oxide; A, Adenine; T, 1,2,4-triazole; BTA, benzotriazole; Pyoxam, (α-pyridyl)oxamide; bipy, bipyridine; H$_2$D, dimethyldioxime; COD, cyclooctadiene; TU, tetramethylthiuram disulfide p. 239; Dtc, dithiocarbamate.

with the complex layer, which hinders the entrance of reagents to the reaction zone. The process goes from the kinetic region into the diffusion region.

Copper and carbon tetrachloride reaction in dimethylacetamide (DMA) gave $Cu_2Cl_3OH(DMA)_2$, which was formed only in the presence of water [71]. The $CuCl_2(DMSO)_2$ complex that was isolated from the Cu–CCl$_4$–DMSO system had DMSO molecules in *trans* positions, and bridging chlorine atoms forming infinite screw chains, as was concluded by X-ray analysis [78].

The interaction of copper powder with C_4H_9Br in DMSO gave two products: $(CuBr_2)_2(DMSO)_3 \cdot H_2O$ and $CuBr_2(DMSO)_3 \cdot H_2O$ [71]. The presence of unsplit bands $\nu(CuO)$ (at about 485 and 490 cm^{-1}, respectively) and $\nu(SO)$ (at about 935 cm^{-1}) in the IR spectra indicates the *trans* structure of the complexes obtained. The authors noted that in the presence of only one halogen atom in the hydrocarbon, the copper is oxidized only in DMSO, while for two and more halogen atoms in the oxidant, the process is possible in both DMSO and DMF.

Interaction of copper with the solutions of PyO–RX (RX = CCl$_4$, CHBr$_3$, 1,2-dibromoethane) leads to the formation of complexes with both Py and PyO as ligands (Table 3.3) [82]. The pyridine complexes are obtained with a small excess of PyO over copper. In the PyO–CHBr$_3$ system the compound $Cu(PyO)_4Br_2$ was obtained at 30°C and with higher PyO contents in the initial mixture, while at lower concentrations and a temperature of 60°C it was not formed. The main products of the oxidation reaction were $Cu(PyO)_2Br_2$ and $Cu(Py)_2Br_2$. Only pyridine complexes $Cu(Py)_2X_2$ and $Cu_4OBr_6(Py)_4$ with bridging oxygen were obtained in Cu–PyO–1,2-dichloro- or dibromoethane systems.

A solution of triphenylphosphine oxide in carbon tetrachloride oxidizes copper very slowly. In the presence of methyl ethyl ketone, depending upon the degree of dryness, the compound $Cu(TPPO)_2Cl_2$ and the tetranuclear complex $Cu_4OCl_6(TPPO)_4$ similar to the ones mentioned above, were isolated [82].

The rate of interaction in Cu–CCl$_4$–Solv systems increases in the presence of an additional ligand. The competing ligand may partly or completely substitute for the solvent molecules in the coordination sphere of the complexes. Oxidation of copper in a DMF–CCl$_4$–adenine (A) system gave the complex $CuCl_2(A)(DMF)$; compounds $CuCl_2T$ and $CuCl_2(DMF)T_2$ were obtained, depending on the component ratio, in the DMF–CCl$_4$–1,2,4-triazole (T) system, and in the DMF–CCl$_4$–benzotriazole (BTA) and DMF–

CCl_4–N,N'-bis(α-pyridyl)oxamide (Pyoxam) systems the compounds $[CuCl_2(BTA)(DMF)]\cdot 2H_2O$ and $[Cu_2Cl_4(Pyoxam)_3]$ were obtained respectively [98]. Analysis of IR spectra at low frequencies revealed end and bridging chlorine atoms in complexes with A, T and Pyoxam, while in BTA complexes only end atoms were found.

Considerable acceleration of copper oxidation was observed upon the addition of ring-forming ligands of dimethylglyoxine (H_2D) and 2,2'-bipyridyl (bipy). $[CuCl_2(H_2D)]_2\cdot DMSO$, $CuCl_2(H_2D)(DMF)$, and $CuCl_2(2,2'$-bipy) compounds are formed in this case [72]. Coordination of DMSO via the oxygen atom ($\nu(CuO) = 480, 500$; $\nu(SO) = 960$ cm^{-1}) and the absence of coordination for DMF were reported by the authors with the help of IR spectra.

An ethanol solution of carbon tetrachloride does not oxidize copper, but in the presence of an equimolar quantity of H_2D, the compound $CuCl_2(H_2D)$ is formed. In the general case [72] the oxidation of copper goes through the formation of copper(I) compounds. This was confirmed by stabilizing the intermediate state of copper with cycloocta-1,5-diene (1,5-COD) and isolating $[Cu(1,5-COD)Cl]_2$ [72]. An interesting example of stepwise copper oxidation and formation of compounds with unusual composition and structure is described in [99]. The behavior of copper powder in the three-component heterophase system with insoluble tetramethylthiuram disulfide (TU) was studied. The latter may lead to the formation of a dithiocarbamate chelate knot with the copper. The copper powder was placed in the solutions of DMSO–CI_4 or DMF–CI_4 in the presence of TU for several days at room temperature. The product was obtained as a black powder. The compound $[Cu_3I_2(Dtc)_5]$, containing two copper(II) atoms with two dithiocarbamate (Dtc) fragments and one copper(III) with one dithiocarbamate was isolated from DMSO. On decreasing the DMSO, $[Cu_2I_4(Dtc)_3]$ was formed, which in the opinion of Kurskov et al. contained copper(II) with one Dtc molecule and copper(III) with two dithiocarbamate fragments. $[Cu_2I(Dtc)_2]$, most probably containing copper(I) and copper(II) atoms, was isolated from DMF. The magnetic measurements showed approximately 40% of copper(II) [99]. The oxidant in the system could be either CI_4, or I_2, as well as its molecular complex $(CH_3)_2S\cdot I_2$, formed in the reaction.

The compounds formed by the interaction of nickel and cobalt powders with nonaqueous solutions of halogenated hydrocarbons are listed in Table 3.4. Their composition depends mainly on the nature of the solvent. The difference between nickel and cobalt behavior is worth mentioning. Cobalt,

TABLE 3.4
Compounds formed in interaction of cobalt and nickel powders with nonaqueous solutions of halogenated hydrocarbons

System	Compound	Reference
Ni–CCl$_4$–DMF	NiCl$_2$·nDMF·2H$_2$O (n = 2, 3)	71
Ni–CCl$_4$–DMAA	NiCl$_2$·DMAA·nH$_2$O (n = 1, 3)	71
Ni–CCl$_4$–DMSO	NiCl$_2$·nDMSO (n = 3, 6)	71
Ni–C$_4$H$_9$Br–DMSO	NiBr$_2$·6DMSO	71
Ni–CH$_3$I–DMSO	NiI$_2$·6DMSO·H$_2$O	71
Ni–CCl$_4$–DMSO–2,2′-bipy	NiCl$_2$·2,2′-bipy	72
Ni–CCl$_4$–DMSO–H$_2$D	NiCl$_2$·DMSO·H$_2$D	72
Ni–CCl$_4$–AN–H$_2$D	Ni(HD)$_2$	72
Ni–CCl$_4$–DMF–H$_2$D	NiCl$_2$·2H$_2$D·DMF	72
Ni–CHBr$_3$–PyO	Ni(PyO)$_6$Br$_2$	82
Ni–CH$_3$I–PyO	Ni(PyO)$_6$I$_2$	82
Co–CCl$_4$–Solv	CoCl$_2$	71
(Solv = ether, AN, ethanol)		
Co–CCl$_4$–DMF–ether	CoCl$_2$·2DMF	71
Co–CCl$_4$–DMA	CoCl$_2$·2DMAA	71
Co–CCl$_4$–DMSO	[Co(DMSO)$_6$][CoCl$_4$]	71
Co–C$_4$H$_9$Br–DMSO	[Co(DMSO)$_6$][CoBr$_4$]	71
Co–CH$_3$I–DMSO	[Co(DMSO)$_6$][CoI$_4$]	71
Co–CCl$_4$–H$_2$D–C$_2$H$_5$OH	H[CoCl$_2$(HD)$_2$]	72
Co–CCl$_4$–PyO	Co(PyO)$_3$Cl$_2$	82
Co–CCl$_4$–TMAO	Co(TMAO)$_2$Cl$_2$	82
Co–CCl$_4$–C$_2$H$_5$OH–Ph$_3$PO	Co(Ph$_3$PO)$_2$Cl$_2$	82
Co–CCl$_4$–Py–Hacac	Co(Py)$_2$(acac)$_2$	106

DMA, dimethylacetamide; TMAO, trimethylamine N-oxide.

unlike nickel, interacts with ether, acetonitrile or ethanolic solutions of carbon tetrachloride, forming cobalt(II) chloride. MCl$_2$·nSolv solvates were isolated from DMF and DMA. In DMSO with carbon tetrachloride both nickel and cobalt are dissolved, but the latter forms compounds of more complex composition (Table 3.4). On introduction of an additional ligand capable of forming stable complexes with the metals, the rate of interaction increases as in the case of copper. Here cobalt and nickel also behave differently. Thus, the interaction of cobalt with carbon tetrachloride and dimethylglyoxine in ethanol gives the compound H[CoCl$_2$(HD)$_2$], which, in the authors' opinion, is a typical acid [72]. The nickel gives simple dimethyl-dioximate from acetonitrile, and respective solvates from DMF and DMSO

(Table 3.4). Kurskov et al. [72] suggested the proton of dimethylglyoxine to be an oxidant in these systems.

The presence of a small quantity of copper powder also increases the rate of nickel interaction with 2,2′-bipy and carbon tetrachloride in DMSO, giving the Ni(2,2′-bipy)$_2$Cl$_2$ complex [72]. Cobalt and nickel react quite readily with acetylacetone solutions of carbon tetrachloride or bromoform in the presence of nitrogen-containing ligands (pyridine, triethylamine, 2,2′-bipy), forming acetylacetonates of various compositions [106]. Thus, cobalt gives cobalt(II) acetylacetonate, cobalt(III) acetylacetonate and mixed complex Co(Py)$_2$(acac)$_2$. Kinetic studies showed that the rate of dissolution of the metal in the systems is determined by its sublimation energy and depends on the concentration of amino-ligand [106].

Similarly to copper, cobalt and nickel interact with pyridine oxide in the presence of halogenated hydrocarbons (carbon tetrachloride, chloroform, bromoform, methyl iodide, 1,2-dichloro- or dibromoethane). Contrary to copper, however, pyridinate complexes are not formed, and only pyridine oxide complexes were obtained (Table 3.4). When triethylamine oxide was used instead of pyridine oxide, the reaction rate decreased [82], which is explained by the low solubility of (TMAO) in halogenated hydrocarbons. A similar complex with triphenylphosphine oxide was obtained only for cobalt. Addition of methanol or ethanol increases the rate of the interaction of cobalt with the solution of triphenylphosphine oxide in carbon tetrachloride [82].

The compositions of the iron compounds obtained on oxidative dissolution of iron with halogenated hydrocarbons are listed in Table 3.5. Their formation was observed for the first time on the interaction of iron powder with carbon tetrachloride in DMSO in the oligomerization of butadiene [109]. Binuclear complexes of Fe$_2$Cl$_4$(DMF)$_3$ and Fe$_2$Cl$_5$(DMSO)$_3$, with iron atoms in different oxidation states and bridging chlorine atoms, were obtained in this case. Later the possibility of obtaining complexes which contain both iron(II) and iron(III) was demonstrated for the interaction of iron powder with carbon tetrachloride in dipolar aprotic solvents. The composition of the complexes depends on both the nature of the reagents and the reaction conditions. Thus, the interaction of iron with carbon tetrachloride in DMSO in vacuo at 100–120°C gives two complexes, namely *cis*-[Fe^{III}Cl$_2$(DMSO)$_4$]Cl and *cis*-[Fe^{III}Cl$_2$(DMSO)$_4$][Fe^{III}Cl$_4$], and in argon at 60–70°C only *trans*-[FeCl$_2$(DMSO)$_4$]Cl is obtained [68,78]. In the presence of benzene at 100°C two complexes were formed regardless of the composition

TABLE 3.5
Compounds formed in interaction of iron powder with nonaqueous solutions of halogenated hydrocarbons

System	Compound	Reference
CCl_4–AN	$[Fe^{II}(AN)_6][Fe^{III}Cl_4]_2$	68, 78
CCl_4–C_6H_5CN	$[Fe^{II}(C_6H_5CN)_6][Fe^{III}Cl_4]_2$	68
CCl_4–DMF	$trans$-$[Fe^{III}(DMF)_4Cl_2][Fe^{III}Cl_4]$	68, 78
	$[Fe^{III}(DMF)_3Cl_2][Fe^{III}Cl_4]$	
CCl_4–DMA	$Fe^{II}Fe^{III}Cl_8(DMA)_6$	107
CCl_4–DMSO	cis-$[Fe^{III}(DMSO)_4Cl_2]Cl$	68, 78
	cis-$[Fe^{III}(DMSO)_4Cl_2]_2[Fe^{II}Cl_4]$	
	$trans$-$[Fe^{III}(DMSO)_4Cl_2][Fe^{III}Cl_4]$	
	$trans$-$[Fe^{III}(DMSO)_4Cl_2]Cl_2$	
CCl_4–C_6H_6–DMSO	$[Fe^{III}(DMSO)_3Cl_2][Fe^{III}Cl_4]$,	78
	$[Fe^{III}(DMSO)_5Cl][Fe^{III}Cl_4]_2$,	
	$[Fe^{III}(DMSO)_5Cl][Fe_2Cl_6O]$	
CCl_4–PyO	$Fe(PyO)_3Cl_3$	74, 82
$CHBr_3$–Ph_3PO	$Fe(Ph_3PO)_3Br_3$	82

of the gaseous phase: $trans$-$[Fe^{III}Cl_2(DMSO)_4]Cl$ and $trans$-$[Fe^{III}Cl_2(DMSO)_4][Fe^{III}Cl_4]$. When the temperature rose above 100°C the formation of cis-complexes became preferred [68,78].

The iron powder interacts rather slowly with Ph_3PO (TPPO) solutions in carbon tetrachloride or bromoform, even in the presence of oxygen. The reaction rate increases on addition of methanol or methyl ethyl ketone [82] and the complex with two Ph_3PO molecules is formed. A complex of similar composition was also obtained with pyridine oxide (Table 3.5).

There are some reports on the direct synthesis of other metal complexes with halogenated hydrocarbons [75]. Thus, titanium metal interacts with carbon tetrachloride in DMA and DMF, but not in DMSO. Kurskov et al. [75] explain this fact by the ability of DMSO to transmit oxygen, forming an oxide film on the metal. A dark violet solution, which turns red after some time, is formed in DMF. The compound $[TiCl(DMF)_5]Cl_3$ was isolated from this solution. $TiCl_4 \cdot 2DMA$ was obtained in the same manner in DMA.

Interaction of chromium with carbon tetrachloride in DMF leads to the formation of chromium(II) complexes with general composition $n(CrCl_2) \cdot mDMF$ [110]. Compounds $CrCl_3 \cdot xROH \cdot yH_2O$ were obtained from alcohol solutions [86]. Chulkevich et al. assume that the oxidation in the system is caused by the interaction products of carbon tetrachloride and alcohol that form on the metal surface.

Silver, gold and even platinoids interact with halogenated hydrocarbons in quite soft conditions [94,96,97,100]. Thus, palladium dissolves in CCl_4–DMA [111] and $CHBr_3$–PyO [112] systems, forming complexes $(HL)_2[Pd_2X_6]$ (L = DMA, PyO; X = Cl, Br). The main product of interaction of palladium with carbon tetraiodide and tetramethylthiuram disulfide in DMSO is $PdI[(CH_3)_2S][(CH_3)_2NCS_2]$ [97,100]. The compound $PdBr_2·2Ph_3PO$ was isolated from a $Pd–Ph_3PO–Hacac–CHBr_3$ system. The complexes $Pt(2,2'-bipy)Br_4$, $Pd(2,2'-bipy)Cl_4$ and $Pd(2,2'-bipy)Br_2$ were obtained on dissolving platinum and palladium in RX-$2,2'$-bipy–DMSO systems [111]. The last of these compounds appears to be a suitable catalyst for dehalogenation. Thus, 3,4-dichlorobutene is converted into butadiene at 180°C with 75% yield in its presence [111]. From the interaction of gold and butyl bromide in DMSO, the complex $[Me_2BuSO][AuBr_4]$ was formed [104,113]. Silver in such systems forms complexes with trimethylsulfonic cations, $[Me_3S]AgBr_2$ and $[Me_3S]Ag_2X_3$ (X = Cl, Br) [114].

Among the attempts to explain the mechanism of interactions in systems with halogenated hydrocarbons [81,82,96,102] the concept of a donor–acceptor electron-transport system, suggesting the formation of dipolar "one" compounds as intermediates [96,102], should be noted. To confirm the mechanism, the interaction of copper with nonaqueous solutions of "one" compounds $[(CH_3)_2N=CHOC_2H_5]BF_4$, $[(CH_3)_2(C_2H_5)S=O]BF_4$, $[(CH_3)S=O]I$ was studied, and the respective complexes of copper(I) and copper(II) were obtained [102].

Experimental procedures

Synthesis of $[CH_3NH_3][Cu(HCOO)_3]$

Copper powder (0.4 g, 0.006 mol), *N*-methylformamide (6 cm^3) and CCl_4 (12 cm^3) were placed in a flask and the mixture was heated to 70°C, refluxed and stirred until total dissolution of the copper was observed (2 h). The green solution was filtered and allowed to stand at room temperature for three days after which long turquoise needles of $[CH_3NH_3][Cu(HCOO)_3]$ separated. These were filtered off, washed with mixture of EtOH and Et$_2$O, and finally dried at room temperature (yield 1.07 g, 72%) [101].

Synthesis of Cu(Ph₃PO)₂Cl₂

Copper powder (0.127 g, 0.002 mol), triphenylphosphine (1.39 g, 0.005 mol), CCl₄ (2 cm³) and dry methyl ethyl ketone (8 cm³) were placed in a flask and the mixture was boiled, refluxed and stirred for 3 h. After slow cooling, yellow crystals of the product precipitated from the solution. These were filtered off and washed with cold methyl ethyl ketone [82].

Synthesis of [Feᴵᴵ(CH₃CN)₆][FeᴵᴵᴵCl₄]₂

Iron powder (1.233 g, 0.022 mol) prepared from Fe(CO)₄, CH₃CN (25 cm³) and CCl₄ (50 cm³) were placed in a flask and the mixture was heated to 70°C, refluxed and stirred under CO₂ until total dissolution of the iron was observed (4 h). The reddish-brown solution was allowed to stand at room temperature for three days, after which the shiny yellow needles of [Feᴵᴵ(CH₃CN)₆][FeᴵᴵᴵCl₄]₂ separated. These were filtered off and washed with CCl₄ and pentane. They were purified by recrystallization from acetonitrile with a mixture of CCl₄ and pentane (10:1) [68].

Synthesis of Ni(PyO)₆Br₂

Nickel powder (0.1 g, 0.002 mol), PyO (1.1 g, 0.012 mol) and CHBr₃ (5 cm³) were placed in a flask and the mixture was heated to 80°C, stirred and refluxed for 20 h. The hot reaction mixture was filtered and cooled slowly. Dark green crystals of Ni(PyO)₆Br₂ were filtered off and washed with Et₂O [82].

Synthesis of [Co(DMSO)₆][CoI₄]

Cobalt powder (0.15 g, 0.0025 mol), DMSO (4 cm³) and CH₃I (0.3 cm³) were placed in a flask and the mixture was heated to 50°C, stirred and refluxed for 5 h. Then acetone (1 cm³) and Et₂O were added dropwise to the solution until a precipitate of [Co(DMSO)₆][CoI₄] was observed to develop. The crystals were filtered off, washed with Et₂O and finally dried in vacuo [71].

3.4 DIRECT SYNTHESIS IN SOLUTIONS OF PROTON-DONOR REAGENTS

Interactions in such systems become possible if a component with active hydrogen is present [115]. It may be either a ligand (HL) or a solvent (HSolv). Therefore two main types of interactions will be considered (reactions (3.6) and (3.7)).

$$M + nL + 2HSolv + \tfrac{1}{2}O_2 = M(Solv)_2 \cdot nL + H_2O \qquad (3.6)$$

$$M + 2HL + n\text{Solv} + \tfrac{1}{2}O_2 = M(L)_2 \cdot n\text{Solv} + H_2O \qquad (3.7)$$

Coordination compounds of copper obtained from the interaction of copper powder with proton-donor reagent solutions are listed in Table 3.6. Systems of the first type are represented by solutions of ethylenediamine (En), 2,2'-bipy, 1,10-phen and Py in alcohols, acetic acid and acetylacetone [72,116,118]. Metallic copper, which is insoluble in acetic acid under normal conditions, does react in the presence of the above ligands. The compounds $Cu(Ac)_2(L) \cdot HAc$ (L = 2,2'-bipy, 1,10-phen, En), and $Cu_2Ac_4 \cdot 2Py$ were isolated in this case [116]. When ring-forming amines are present, the acidic properties of the alcohol strengthen enough to form copper alcoholate complexes (Table 3.6). Metallic copper does not react with acetylacetone, but in the presence of 2,2'-bipy the compound $Cu(acac)_2$ is obtained [72].

TABLE 3.6
Compounds formed from interaction of copper with solutions of proton-donor reagents

Solution	Compound	Reference
HAc–L (L = 2,2'-bipy, 1,10-phen, En)	$Cu(Ac)_2(L) \cdot HAc$	116
HAc–Py	$Cu_2(Ac)_4 \cdot 2Py$	116
ROH–L (L = 2,2'-bipy, 1,10-phen, En; ROH = methanol, ethanol)	$Cu(OR)_2(L) \cdot nROH$	117, 118
Hacac–2,2'-bipy	$Cu(acac)_2$	72
H_2O–1,10-phen	$Cu_2(CO_3)_2(1,10\text{-phen})_3 \cdot 11H_2O$	95
H_2O–amide–L (amide = MA, DMA, FA; L = 2,2'-bipy, 1,10-phen)	$Cu_3(OH)_2(CO_3)_2(L)_3 \cdot H_2O$, $Cu(HCOO)(CN)(2,2'\text{-bipy}) \cdot xH_2O$, $Cu(HCOO)_2 \cdot 2FA$	84
C_2H_5OH–H_2D	$Cu(HD)_2$	72
C_2H_5OH–imidazole (HL)	$Cu(L)_2$	119
Solv–HSal (Solv = hexane, hexadecane, dioxane)	$Cu(Sal)_2$	120
Salicylaldoxime (HL)–Solv (Solv = anisole, phenetole, AN, benzaldehyde)	$Cu(L)_2$	112
ROH–HEa, (ROH = methanol, ethanol)	$Cu(Ea)_2 \cdot ROH$, $Cu(Ea)_2(HEa)$	121

En, ethylenediamine; HSal, salicylalaniline; HEa, ethanolamine.

Rapid reaction of copper with nitromethane, nitroethane and phenylnitromethane in the presence of pyridine, imidazole, N-phenylimidazole and other nitrogen-containing heterocycles is explained by the action of an active proton in the nitro compound [112]. At the same time the nature of these complexes and their mechanism of formation are not known. Copper metal dissolves in water in the presence of 1,10-phen. In this case the hydroxo complex of copper(I) is obtained in the absence of atmospheric CO_2, and the respective carbonate, $Cu_2(CO_3)(1,10\text{-phen})\cdot 11H_2O$, is formed by the action of CO_2 [95].

Reactions of copper with aqueous solutions of amides and their derivatives in the presence of ring-forming ligands (2,2'-bipy, 1,10-phen) have been studied [84]. Thus, the polynuclear carbonate complexes $Cu_3(OH)_2(CO_3)_2L_3\cdot H_2O$ and $Cu(HCOO)CN(2,2'\text{-bipy})\cdot xH_2O$ were formed by oxidizing copper in MA(DMA, FA)–water–L (L = 2,2'-bipy, 1,10-phen) systems. The carbonate anion of the first compound appears upon hydrolysis of amides followed by the oxidation of acetic acid to carbonic acid, while cyanide anions form during formamide decomposition.

Some chelating ligands (dimethylglyoxine, imidazole, monoethanolamine, salicylalaniline and others) have proton-donor properties in solution [72]. Metallic copper interacts with solutions of dimethylglyoxine in ethanol, anisole, and phenetole, forming copper dimethyldioximate. When a copper plate was put into an ethanolic solution of imidazole (HL) CuL_2 was formed on the surface [119]. The authors suggested that copper coordinates with the pyridine nitrogen atom, forming an unstable intermediate, which under the action of atmospheric oxygen is converted into the final product, losing a water molecule.

Systems of the second type may include copper interactions with salicylalaniline (HSal) in hexane, hexadecane or dioxane, forming $Cu(Sal)_2$ [120]. The kinetics of the interaction of copper powder with HSal in DMF at 20–50°C has been studied [103]; the activation energy was found to be 23 $kJ\,mol^{-1}$. Khentov et al. [103] suppose that the dissolution of the copper is preceded by its surface oxidation with atmospheric oxygen, dissolved in an organic solvent, the interaction of the ligand and oxide being a rate-restricting step.

When *syn-* and *anti-*oximes of benzoic and salicylic aldehydes are boiled with copper in anisole, phenetole, acetonitrile or benzaldehyde, Beckman regrouping occurs accompanied by copper dissolution and formation of the respective copper complexes [112].

Oxidation of copper with monoethanolamine in alcohol (methanol, ethanol) solutions leads to octahedral complexes, solvated either by alcohols or HEa [121]. The composition of the complexes depends on the monoethanolamine content in the mixture. The complexes solvated with alcohol are obtained at low concentrations, and at higher HEa content the substitution of alcohol by ethanolamine is observed in the product (reactions (3.8) and (3.9)).

$$Cu + 2HEa + ROH + \tfrac{1}{2}O_2 = Cu(Ea)_2 \cdot ROH + H_2O \qquad (3.8)$$

$$Cu + 3HEa + \tfrac{1}{2}O_2 = Cu(Ea)_2 \cdot (HEa) + H_2O \qquad (3.9)$$

The same types of reactions can occur in the interaction of copper, silver, gold, rhodium and palladium with nonaqueous solutions of hydrohalogenic acids [81,104,113,122,123].

An interesting set of conversions accompanying metal dissolution is considered in the process of DMSO condensation with ketones in the presence of silver complexes (reaction (3.10)) [124].

$$Me_2SO + R'CH_2COR \xrightarrow[-R'OH]{HAgX_2} [Me_2S^\pm CH_2COR][AgX_2]^-$$

$$X = Br, I \qquad (3.10)$$

Data on other metal interactions with proton-donor reagents are not numerous (Table 3.7). Thus, the interaction of iron with acetylacetone leads to iron(III) acetylacetonate. The rate of the reaction is influenced appreciably by atmospheric oxygen. It was stated [125] that iron(II) acetylacetonate is an intermediate, which catalyzes oxidation of some acetylacetone to CO and CO_2. This system was studied more fundamentally, and the represented by reactions (3.11) mechanism was suggested for the interaction:

$$Hacac \rightarrow H^+ + acac^-$$
$$Fe + 2H^+ \rightarrow Fe^{2+} + H_2$$
$$Fe^{2+} + 2acac^- \rightarrow Fe(acac)_2 \qquad\qquad (3.11)$$
$$Fe(acac)_2 + acac^- \rightarrow Fe(acac)_3 + e^-(?)$$

Other metals, for example, nickel, cobalt, chromium, tin and titanium, also dissolve in acetylacetone in the presence of oxygen (reaction (3.12)) [131].

TABLE 3.7
Compounds formed on interaction of cobalt, nickel and iron powders with solutions of proton-donor reagents

System	Compound	Reference
Fe–Hacac	Fe(acac)$_3$	125, 126
	Fe(acac)$_2$	125
Fe–HL–Solv	Fe(L)$_2$	112
(HL = salicylaldoxime; Solv = acetone, phenetole, AN)		
Fe–HL (HL = trifluoroacetylacetone)	Fe(L)$_2$	72
Co–2,2′-bipy–Hacac	Co(acac)$_2$·2,2′-bipy	72
Co–Hoxine–DMF	Co(oxine)$_2$·2H$_2$O	72
Co–L–ROH	Co(OR)$_2$(L)·nROH	128
(L = 2,2′bipy, 1,10-phen; ROH = CH$_3$OH, C$_2$H$_5$OH)		
Co–L–HAc (L = 2,2′-bipy, 1,10-phen)	Co(Ac)$_2$(L)·HAc	129
Ni–2,2′-bipy–Hacac	Ni(acac)$_2$·2,2′-bipy	72
Ni–HL–Solv	Ni(L)$_2$	121
(HL = salicylaldoxime; Solv = acetone, phenetole, AN)		
Ni–H$_2$D–Solv (Solv = anisole, phenetole)	Ni(HD)$_2$	130
Ni–L–ROH	Ni(OR)$_2$(L)·nROH	128
(L = 2,2′-bipy, 1,10-phen; ROH = CH$_3$OH, C$_2$H$_5$OH)		
Ni–L–HAc (L = 2,2′-bipy, 1,10-phen)	Ni(Ac)$_2$(L)·HAc	129

$$M + 2Hacac + \tfrac{1}{2}O_2 \rightarrow M(acac)_2 + H_2O \tag{3.12}$$

More active metals, for example, magnesium and zinc, react with acetylacetone even in an inert atmosphere (reaction (31.3)).

$$M + 2Hacac \rightarrow M(acac)_2 + H_2 \tag{3.13}$$

The interaction of yttrium with nonaqueous solutions (acetone, AN, DMF, DMSO, dioxane, ether) of acetylacetone was studied recently, and formation of Y(acac)$_3$·2H$_2$O and Y(acac)$_3$·nH$_2$O·mSolv complexes was reported

[132]. Complexes with these ligands were isolated in the presence of amines (Py, 2,2′-bipy) [132].

Nevertheless, data on metal interaction with acetylacetone are somewhat contradictory. Thus, Kurskov et al. [72] report that cobalt and nickel do not react with acetylacetone. The discrepancy may be explained by the difference in the purity grade of the reagent, since the presence of water may influence its acidic properties. In the presence of 2,2′- or 4,4′-bipy (L), cobalt and nickel dissolve in acetylacetone forming mixed complexes M(acac)$_2$·L [72]. The interaction rate for 2,2′-bipy is almost twice that for 4,4′-bipy, in view of the different stability of the complexes. Fluorinated β-diketones have stronger acidic properties and interact with metals even in an inert atmosphere [133]. The reactions are catalyzed by small quantities of water. The interaction of iron and chromium with trifluoroacetylacetone is used in the rapid analysis of the alloys [127].

Metal β-diketonates were obtained by oxidation of metal powders with acetylacetone in the presence of oximes (benzaldoxime, acetophenoxime, acetoxime) in 40–90% yield [112].

Metallic cobalt reacts with 8-oxyquinoline in DMF at 60°C to form a brown solution, from which compounds Co(oxine)$_2$ [111] and Co(oxine)$_2$·2H$_2$O [72] were isolated. Nickel dimethyldioximate was obtained upon nickel powder interaction with dimethylglyoxine in anisole or phenetole in nearly 100% yield [130]. In the same manner chelate complexes with other ligands were obtained, for example, 3-(o-hydroxyphenyl)-5-methyl(phenyl, styryl)-1,2,4-oxadiazole, 1-phenyl-3-methyl-5-(o-hydroxyphenyl)-1,2,4-triazole, in yields up to 70% [112]. Similarly to copper, cobalt and nickel interact with alcohol solutions of 2,2′-bipy and 1,10-phen, forming compounds of the same nature, but the rate of interaction is low [128]. Cobalt and nickel metals dissolve slowly in acetic acid, but the rate of the reaction becomes considerably greater in the presence of 2,2′-bipy or 1,10-phen. Acetate complexes containing ligands and acetic acid were isolated from the solutions (Table 3.7). Pyridine and ethylenediamine do not influence the dissolution rate of cobalt and nickel in acetic acid. In this case only simple metal acetates were obtained, in contrast to the case of copper products [129].

Bacterial dissolution of metals is due to interaction of the metal with metabolic products containing aminoacids. It was shown [134] that transition metal interaction with aminoacids and peptides is most effective in polar solvents.

Experimental procedures

Synthesis of Ni(acac)$_2$·2,2'-bipy

Nickel powder (0.20 g, 0.0034 mol), 2,2'-bipy (0.54 g, 0.0034 mol) and acetylace-tone (10 cm^3) were placed in a flask and the mixture was heated to 60°C, stirred and refluxed for 8 h. After cooling, beige crystals of the product precipitated from the solution. These were filtered off, washed with acetone and dried [72].

Synthesis of Co(oxine)$_2$·2H$_2$O

Cobalt powder (0.10 g, 0.0017 mol), 8-hydroxyquinoline (0.48 g, 0.0033 mol) and DMF (5 cm^3) were placed in a flask and the mixture was heated to 60°C, stirred and refluxed for 8 h. A brown residue of Co(oxine)$_2$·2H$_2$O was filtered off, washed with Et$_2$O and EtOH and dried [72].

Synthesis of Cu(HD)$_2$

Copper powder (0.212 g, 0.0033 mol), dimethylglyoxime (0.773 g, 0.0066 mol) and CH$_3$OH (10 cm^3) were placed in a flask and the mixture was heated to 40°C, stirred and refluxed for 5 h. The solution was filtered. After cooling, dark violet crystals of the product precipitated from the solution. These were filtered off, washed with EtOH and dried [72].

3.5 DIRECT SYNTHESIS WITH SOLUTIONS OF AMMONIUM SALTS

Ammonium salts attract attention as non-aggressive reagents in inorganic synthesis. Their application in thermal synthesis of anhydrous metal halides [135–141] based on reaction (3.14) is well known.

$$MO + 2NH_4X = MX_2 + 2NH_3 + H_2O \qquad (3.14)$$

The same reactions are known for metal carbonates [135–138] and for metals with low melting points [139–141]. Application of nonaqueous solvents con-siderably increases the possibilities of obtaining coordination compounds with the help of direct synthesis. The composition of the complex is mainly determined by the nature of the metal, the ammonium salt anion and the solvent, as well as by their ratio in the mixture.

The products of metal powder or metal oxide interaction with ammonium salt solutions can be neutral salts [58,142–147], and hydroxy salts, for cad-mium [44] and lead [58,143] in particular.

3.5.1 Complexes containing ammonia and solvent molecules

Direct synthesis gives a route to one-step isolation of three main types of products: ammines ($M(NH_3)_nX_2$), solvates ($M(Solv)_nX_2$), and amminosolvates ($M(NH_3)_n(Solv)_mX_2$). Examples of the compounds obtained from metal powders or metal oxides and ammonium salts are listed in Table 3.8. Ammines are formed preferably in alcohol and acetonitrile solutions (reaction (3.15)).

TABLE 3.8
Composition of amine complexes formed upon interaction of metal powders and metal oxides with nonaqueous solutions of ammonium salts

Initial system	Compound	Reference
$Zn(ZnO)-NH_4X-Solv$ (Solv = ROH, AN, DMF; X = Cl, Br, I)	$Zn(NH_3)_2X_2$	42, 45, 52, 148–151
$Zn(ZnO)-NH_4I-Solv$ (Solv = ROH, AN, DMF, DMSO)	$Zn(NH_3)_2I_2$	52, 55, 150–153
$Cd(CdO)-NH_4Cl-Solv$ (Solv = ROH, AN, DMF, DMSO)	$Cd(NH_3)_2Cl_2$	152
$Cd(CdO)-NH_4X-Solv$ (Solv = ROH, AN, DMF; X = Br, I)	$Cd(NH_3)_2X_2$	151
$Cd(CdO)-NH_4SCN-Solv$ (Solv = ROH, AN)	$Cd(NH_3)_2(SCN)_2$	44, 45
$Cu_2O-NH_4SCN-AN$	$Cu(NH_3)_2(SCN)_2$	153, 154
$Ni-NH_4Cl-DMSO$	$Ni(NH_3)_6Cl_2$	67, 155
$Ni-NH_4Br-Solv$ (Solv = n-propanol, AN, DMF)	$Ni(NH_3)_4Br_2 \cdot H_2O$	156
$Ni-NH_4SCN-Solv$ (Solv = ROH, CH_3CN)	$Ni(NH_3)_2(SCN)_2$ $Ni(NH_3)_4(SCN)_2 \cdot 3H_2O$	51
$Co-NH_4Cl-Solv$ (Solv = n-propanol, AN)	$Co(NH_3)_6Cl_3$	156
$Co-NH_4Br-Solv$ (Solv = ROH, AN)	$Co(NH_3)_6Br_3, Co(NH_3)_2Br_2$	156
$Co-NH_4SCN-AN$	$Co(NH_3)_2(SCN)_2 \cdot (H_2O)_2$ $[Co(NH_3)_5(SCN)](SCN)_2$	48

$$MO + nNH_4X = M(NH_3)_n X_2 + (n-2)HX + H_2O \qquad (3.15)$$

Copper [153,154], zinc [42,45,52,148–151], and cadmium [44,55,151,152] mainly form diammine compounds. An excess of ammonium salt does not influence the product composition for copper, but increases the reaction rate. Thus, when the molar ratio Cu/NH_4Br was increased from 1:2 to 1:5–6, the synthesis time was shortened from 3 to 1 h. Nickel and cobalt may form complexes containing up to six ammonia molecules with an excess of ammonium salt [48,51,67,155,156] (Table 3.8). Thus, the interaction of nickel with ammonium bromide in acetonitrile, DMF or n-propanol (reaction 3.16)) gives tetrammine [156].

$$Ni + 4NH_4Br + \tfrac{1}{2}O_2 = Ni(NH_3)_4 Br_2 \cdot H_2O + 2HBr \qquad (3.16)$$

The interaction of cobalt is complicated, with simultaneous formation of cobalt(II) and cobalt(III) complexes, depending on the nature of the ammonium salt anion. Only one compound, namely $[Co(NH_3)_6]Cl_3$, was obtained with ammonium chloride in acetonitrile or n-propanol. The cobalt interaction with ammonium bromide in acetonitrile or alcohols gives two complexes (reaction (3.17)).

$$3Co + 14NH_4Br + 2O_2 = 2Co(NH_3)_6 Br_2 + Co(NH_3)_2 Br_2$$
$$+ 6HBr + 4H_2O$$

$$(3.17)$$

Pure $Co(NH_3)_2Br_2$ complex could be obtained only from acetonitrile solution, but two complexes were also formed with ammonium thiocyanate under the same conditions (reaction (3.18)).

$$2Co + 7NH_4SCN + \tfrac{5}{4}O_2 = [Co(NH_3)_5(SCN)](SCN)_2$$
$$+ Co(NH_3)_2(SCN)_2(H_2O)_2 + \tfrac{1}{2}H_2O + 2HSCN$$

$$(3.18)$$

As a rule, when a solvent has high solvability (DMF, DMSO), ammonia is wholly substituted with the formation of the corresponding solvates reaction (3.19)); (Table 3.9). The number of solvent molecules per solvate molecule varies from two for lead [58], zinc [52,149], and cadmium [55,152], up to four for cobalt [43,48] and nickel [51]. Zinc and lead form solvate complexes only

TABLE 3.9

Composition of solvates and amminosolvates formed on interaction of metal powders and metal oxides with nonaqueous solutions of ammonium salts

Initial system	Compound	Reference
$Zn(ZnO)–NH_4X–DMSO$ $(X = Cl, Br)$	$Zn(DMSO)_2X_2$	52, 149
$Cd(CdO)–NH_4X–DMSO$ $(X = Br, I)$	$Cd(DMSO)_2X_2$	152
$Cd(CdO)–NH_4SCN–Solv$ $(Solv = DMF, DMSO)$	$Cd(Solv)_2(SCN)_2$	55
$Co–NH_4SCN–DMF$	$Co(DMF)_4(SCN)_2$	52, 151
$Ni–NH_4SCN–Solv$ $(Solv = DMF, DMSO)$	$Ni(Solv)_4(SCN)_2$	51
$Co–NH_4Cl–Solv$ $(Solv = DMF, CH_3OH)$	$Co(Solv)_2(NH_3)_2Cl_2$	149
$Co–NH_4Cl–DMSO$	$Co(NH_3)_4Cl_2·2DMSO$	149
$Ni–NH_4Br–CH_3OH$	$Ni(NH_3)_4Br_2·4CH_3OH$	156
$Ni–NH_4I–CH_3OH$	$Ni(NH_3)_2I_2·2CH_3OH$	156
$PbO–NH_4X–DMSO$ $(X = Br, SCN)$	$Pb(DMSO)_2X$	58

in DMSO, while cadmium can form such complexes with both DMSO and DMF.

$$MO + 2NH_4X + n\text{Solv} = M(\text{Solv})_nX_2 + H_2O + 2NH_3 \qquad (3.19)$$

$$MO + 2NH_4X + 2\text{Solv} = M(\text{Solv})_2X_2 + 2NH_3 + 2H_2O \qquad (3.20)$$

Cobalt and nickel can form mixed complexes (reaction (3.21); Table 3.8). Complexes $Ni(NH_3)_nX_2·mCH_3OH$ were obtained from methanolic solutions of ammonium bromide and iodide. Cobalt interacts with ammonium chloride in methanol or DMF, forming pink crystals of $Co(NH_3)_2(\text{Solv})_2Cl_2$ $(\text{Solv} = DMF, CH_3OH)$ (reaction (3.22)).

$$M + nNH_4X + m\text{Solv} + \tfrac{1}{2}O_2 = M(NH_3)_n(\text{Solv})_mX_2$$
$$+ (n - 2)HX + H_2O \qquad (3.21)$$

$$Co + 2NH_4Cl + \tfrac{1}{2}O_2 + 2Solv = Co(NH_3)_2(Solv)_2Cl_2] + H_2O$$

$$(3.22)$$

Cobalt gives complexes with three or five molecules of a solvent in the interaction with ammonium thiocyanate or bromide in DMSO (reactions (3.23) and (3.24)).

$$Co + 2NH_4SCN + 3DMSO + \tfrac{1}{2}O_2$$
$$= Co(NH_3)(DMSO)_3(SCN)_2 + NH_3 + H_2O$$
$$(3.23)$$

$$Co + 2NH_4Br + 5DMSO + \tfrac{1}{2}O_2$$
$$= [Co(NH_3)(DMSO)_5]Br_2 \cdot H_2O + NH_3$$
$$(3.24)$$

Experimental procedures

Synthesis of Cu(NH₃)₂(SCN)₂

Cu_2O (0.31 g, 0.0022 mol), NH_4SCN (0.84 g, 0.011 mol) and CH_3CN (25 cm^3) were placed in a flask and the mixture was heated to 60°C and stirred until total dissolution of the copper oxide was observed (30 min). In the course of the reaction blue crystals of the product precipitated. These were filtered off, washed with dry isopropanol, and finally dried in vacuo at room temperature (yield 2.13 g; 92%) [153,154].

Synthesis of Zn(NH₃)₂(SCN)₂

ZnO (0.4 g, 0.005 mol) was placed in a flask, a solution of NH_4SCN (0.8 g, 0.01 mol) in CH_3CN (25 cm^3) was added and the mixture was heated to 60°C and stirred until total dissolution of zinc oxide was observed (5 min). Evaporation of the CH_3CN under reduced pressure yielded colorless crystals of $Zn(NH_3)_2(SCN)_2$. These were filtered off, washed with isopropanol, and finally dried in vacuo at room temperature (yield 0.95 g; 89%) [148].

Synthesis of Cd(DMF)₂(SCN)₂

CdO (0.64 g, 0.005 mol), NH_4SCN (0.76 g, 0.01 mol) and DMF (20 cm^3) were placed in a flask and the mixture was heated to 60°C and stirred until total dissolution of cadmium oxide was observed (3 h). Evaporation of the DMF under reduced pressure resulted in the formation of a colorless oil from which crystals of $Cd(DMF)_2(SCN)_2$ precipitated. These were filtered off, washed with isopropanol and Et$_2$O, and finally dried at room temperature (yield 1.12 g; 60%) [55].

3.5.2 Compounds containing complex anions

An excess of ammonium, tetraalkylammonium or alkaline metal salt can lead to the formation of anionic complexes (Table 3.10), as well as to an increase in the interaction rate. Thus, with a three- to four-fold excess of iodide or thiocyanate, zinc and cadmium oxides form tetraanionic and mixed ammino–anionic complexes (reactions (3.25), where $M = K$, $(CH_3)_4N$, $(C_2H_5)_4N$, and (3.26)), and when the ratio becomes 1:6, ammonia substitution occurs with the formation of hexaanionic complexes (reaction (3.27), where $M = Zn, Cd$).

$$CdO + 2NH_4SCN + MSCN = M[Cd(SCN)_3NH_3] + NH_3 + H_2O \tag{3.25}$$

$$ZnO + 4NH_4SCN = (NH_4)_2[Zn(SCN)_4(NH_3)_2] + H_2O \tag{3.26}$$

$$MO + 6NH_4SCN = (NH_4)_4[M(SCN)_6] + 2NH_3 + H_2O \tag{3.27}$$

TABLE 3.10
Composition of anionic complexes formed upon interaction of metal powders and metal oxides with nonaqueous solutions of ammonium salts

Initial system	Compound	Reference
Zn(ZnO)–4NH$_4$SCN–Solv	(NH$_4$)$_2$[Zn(SCN)$_4$(NH$_3$)$_2$]	157
Zn(ZnO)–6NH$_4$SCN–Solv	(NH$_4$)$_4$[Zn(SCN)L]	157
Cd(CdO)–3NH$_4$SCN–Solv	NH$_4$[Cd(SCN)$_3$(NH$_3$)]	44, 55
Cd(CdO)–4NH$_4$SCN–Solv	(NH$_4$)$_2$[Cd(SCN)$_4$]	44, 55
Cd(CdO)–6NH$_4$SCN–Solv	(NH$_4$)$_4$[Cd(SCN)$_6$]	44, 55
Cd(CdO)–2NH$_4$SCN–KSCN–Solv	K[Cd(SCN)$_3$(NH$_3$)]	55, 157
Cd(CdO)–2NH$_4$X–2MX–Solv (X = I, SCN; M = K, (CH$_3$)$_4$N, (C$_2$H$_5$)$_4$N)	M$_2$[CdX$_4$]	55, 157
Cd(CdO)–NH$_4$I–MSCN–Solv (M = K, (CH$_3$)$_4$N)	M[Cd(SCN)$_2$I(NH$_3$)]	157
Cd(CdO)–NH$_4$I–2MSCN–Solv (M = K, (CH$_3$)$_4$N, (C$_2$H$_5$)$_4$N)	M$_2$[Cd(SCN)$_2$I$_2$]	157, 158

The formation of mixed compounds was shown for cadmium oxide. Two types of iodide–thiocyanate complexes were obtained (reactions (3.28) and (3.29), where $M = K, (CH_3)_4N, (C_2H_5)_4N$). The mixed chloride and bromide complexes were not isolated, since during the exchange reaction insoluble $Cd(NH_3)_2X_2$ (X = Cl, Br) complexes were formed.

$$CdO + 2NH_4SCN + MI = M[Cd(SCN)_2I(NH_3)] + H_2O + NH_3$$
$$(3.28)$$

$$CdO + 2NH_4SCN + 2MI = M_2[Cd(SCN)_2I_2] + H_2O + 2NH_3$$
$$(3.29)$$

Experimental procedures

Synthesis of $[(C_2H_5)_4N][Cd(SCN)_3]$

CdO (0.64 g, 0.005 mol), NH_4SCN (0.76 g, 0.01 mol), $(C_2H_5)_4NSCN$ (0.94 g, 0.01 mol) and CH_3OH (20 cm^3) were placed in a flask and the mixture was heated to 60°C and stirred until total dissolution of the cadmium oxide was observed (60 min). Evaporation of the methanol under reduced pressure yielded colorless crystals of the product. These were filtered off, washed with isopropanol, and finally dried in vacuo at room temperature (yield 1.87 g; 90%).

Synthesis of $[(C_2H_5)_4N]_2[CdI_4]$

CdO (0.64 g, 0.005 mol), NH_4I (1.45 g, 0.01 mol), $(C_2H_5)_4NI$ (2.57 g, 0.01 mol) and CH_3OH (20 cm^3) were placed in a flask and the mixture was heated to 60°C and stirred until total dissolution of the cadmium oxide was observed (30 min). Evaporation of the methanol under reduced pressure yielded colorless crystals of the product. These were filtered off, washed with isopropanol, and finally dried in vacuo at room temperature (yield 4.13 g; 94%) [55].

Synthesis of $K_2[Cd(SCN)_2I_2]$

CdO (0.64 g, 0.005 mol), NH_4I (1.45 g, 0.01 mol), KSCN (0.98 g, 0.01 mol) and CH_3OH (20 cm^3) were placed in a flask and the mixture was heated to 60°C and stirred until total dissolution of the cadmium oxide was observed (60 min). Evaporation of the methanol under reduced pressure yielded colorless crystals of the product. These were filtered off, washed with dry isopropanol, and finally dried in vacuo at room temperature (yield 2.57 g; 92%) [158].

3.5.3 Complexes with pyridine

The influence of pyridine on the interaction of zinc and cadmium with ammonium thiocyanates in nonaqueous solutions was studied [159]. It was shown that the character of the solubility curves was identical for the metal and its oxide. Hence, the composition of the complexes obtained is the same. The process is determined by the difference in solubility of the compounds formed in pure pyridine and in its mixtures with other solvents [159].

Table 3.11 lists pyridine complexes obtained in the interaction of metal powders and metal oxides with nonaqueous solutions of ammonium salts and pyridine: it shows that pyridine readily displaces ammonia in the metal coordination sphere (reaction 3.30)).

$$MO + 2NH_4X + nPy = M(Py)_nX_2 + 2NH_3 + H_2O \qquad (3.30)$$

TABLE 3.11

Composition of compounds formed upon interaction of metal powders and metal oxides with solutions of ammonium salts in the presence of pyridine

Initial system	Compound	Reference
Co–NH$_4$X–Py–Solv (Solv = ROH, AN, DMF, DMSO; X = Cl, Br, I, SCN)	Co(Py)$_4$X$_2$	160
Ni–NH$_4$X–Py–Solv (Solv = ROH, AN, DMF, DMSO; X = Cl, Br, I, SCN)	Ni(Py)$_4$X$_2$	160
Zn(ZnO)–NH$_4$X–Py–Solv (Solv = ROH, AN, DMF, DMSO; X = Cl, Br, I, SCN)	Zn(Py)$_n$X$_2$ ($n = 2, 4$)	151, 154, 159
Cd(CdO)–NH$_4$X–Py–Solv (Solv = ROH, AN, DMF, DMSO; X = Cl, Br, I, SCN)	Cd(Py)$_n$X$_2$ ($n = 2, 4$)	151, 154, 159
Cu–NH$_4$X–Py (X = Cl, Br, I, SCN)	Cu(Py)X	161
Cu–NH$_4$X–Py (X = Cl, Br, SCN)	Cu(Py)$_2$X$_2$	161
Cu–NH$_4$I–Py	Cu(Py)$_n$I$_2$ ($n = 4, 6$)	161

The complex composition is not influenced, as a rule, by the nature of the solvent. Therefore, pyridine was used both as the ligand and the solvent. It appeared to be preferable due to the high solubility of complexes in hot pyridine, and their easy crystallization on cooling of the reaction mixture.

The interaction of copper powder with ammonium salts in pyridine has some peculiarities. Thus, copper dissolves in the system when there is free oxygen access, forming yellow or orange solutions which become green after some time. The products of the reaction with ammonium chloride or bromide precipitate on cooling, while iodide and thiocyanate complexes educts are obtained after some hours. The desalting of the solutions in the last case gives copper(I) complexes CuPyX, regardless of the ratio of reagents, but crystals that form upon standing appear to be copper(II) complexes $Cu(Py)_nX_2$ (Table 3.11). Then, during the interaction of copper with ammonium iodide or thiocyanate in pyridine complexes of copper(I) are formed at first (reaction (3.31)), which are then oxidized by atmospheric oxygen to form copper(II) complexes (reaction (3.32)).

$$2Cu + 2NH_4X + 2Py + \tfrac{1}{2}O_2 = 2Cu(Py)X + 2NH_3 + H_2O$$

$$(3.31)$$

$$2Cu(Py)X + 2NH_4X + 2(n-1)Py + \tfrac{1}{2}O_2 = 2Cu(Py)_nX_2 \\ + 2NH_3 + H_2O$$

$$(3.32)$$

The iodide complexes of composition $Cu(Py)_nI_2$ ($n = 4$, 6) are rather unstable and lose pyridine even on washing with ether. Complexes with two, three or five pyridine molecules can be obtained similarly. These compounds are gradually converted into colorless CuPyI on standing.

Experimental procedures

Synthesis of Cu(Py)₂Br₂

Copper powder (0.63 g, 0.01 mol), NH_4Br (1.96 g, 0.02 mol) and pyridine (25 cm^3) were placed in a flask and the mixture was heated to 60°C and stirred until total dissolution of the copper was observed (60 min). After cooling, green crystals of the product precipitated from the solution. Those were filtered off, washed with dry Et$_2$O and finally dried in vacuo at room temperature (yield 3.21 g; 84%) [161].

Synthesis of Zn(Py)₂(SCN)₂

ZnO (0.4 g, 0.005 mol), NH₄SCN (0.8 g, 0.01 mol), CH₃CN (25 cm³) and pyridine (2 cm³) were placed in a flask in the above order, and the mixture was heated to 60°C and stirred until total dissolution of the zinc oxide was observed (5 min). The solution was filtered and half of the acetonitrile was evaporated under reduced pressure. This procedure led to the precipitation of colorless crystals of Zn(Py)₂(SCN)₂. Those were filtered off, washed with dry Et₂O, and finally dried in vacuo at room temperature (yield 1.58 g; 93%) [154].

Synthesis of Cd(Py)₂(SCN)₂

CdO (0.64 g, 0.005 mol), NH₄SCN (0.8 g, 0.01 mol), CH₃CN (25 cm³) and pyridine (1 cm³) were placed in a flask in the above order, and the mixture was heated to 60°C and stirred until total dissolution of the cadmium oxide was observed (30 min). The solution was filtered through a medium-porosity frit, and half of the acetonitrile was evaporated under reduced pressure. This procedure led to the precipitation of colorless crystals of Cd(Py)₂(SCN)₂. Those were filtered off, washed with dry Et₂O and finally dried in vacuo at room temperature (yield 1.74 g; 90%) [151,154].

3.5.4 Complexes with ethylenediamine and its derivatives

The types of compounds formed in the presence of ethylenediamine (En) or its derivatives (diethylenediamine, triethylenediamine) are listed in Table 3.12. Three main types of coordination compounds, namely $M(En)_nX_2$, $M(En)_nX_2 \cdot mSolv$ and $M(NH_3)_2(En)_2X_2$ can be formed. Compound compositions are determined by the ratio of M(MO)/En, and the nature of the solvent. When M(MO)/En = 1:1, the monoethylenediamine complexes are commonly formed, but in some cases (methanolic solutions of ammonium thiocyanate for copper, solutions of ammonium chloride and bromide in methanol and DMF for nickel, and solutions of ammonium chloride in DMF for zinc) the formation of bis(ethylenediamine) compounds was observed (reaction (3.33) and (3.34)).

$$2M + 4NH_4X + 2En + O_2 = M(En)_2X_2 + M(NH_3)_2X_2$$
$$+ 2NH_3 + 2H_2O$$
$$M = Cu, Zn; X = Cl, SCN \tag{3.33}$$

TABLE 3.12

Composition of compounds formed upon interaction of metal powders and metal oxides with solutions of ammonium salts, ethylenediamine and its derivatives

Initial system	Compound	Reference
Cu–NH$_4$X–En–Solv (Solv = ROH, AN, DMF, DMSO; X = Cl, Br, I, NO$_3$, Ac, SCN)	Cu(En)$_n$X$_2$ (n = 1–3)	162, 163
Cu–NH$_4$X–En–Solv (Solv = CH$_3$OH, DMF, DMSO; X = Cl, I)	Cu(En)$_2$X$_2$·mSolv (m = 0.33, 1)	162, 163
Zn(ZnO)–NH$_4$X–En–Solv (Solv = ROH, AN, DMF, DMSO; X = Cl, Br, I, SCN)	Zn(En)$_n$X$_2$ (n = 2, 3)	149, 154, 161
Zn(ZnO)–NH$_4$X–En–Solv (Solv = CH$_3$OH, DMSO; X = Cl, I)	Zn(En)$_2$X$_2$·nSolv (n = 1, 2)	149, 154, 161
Cd(CdO)–NH$_4$X–En–Solv (Solv = ROH, AN, DMF, DMSO; X = Cl, Br, I)	Cd(En)$_n$X$_2$ (n = 1–3)	161
Co–NH$_4$X–En–Solv (Solv = ROH, AN, DMF, DMSO; X = Cl, Br, I, SCN)	Co(En)$_2$X$_2$	164, 165
Ni–NH$_4$X–En–Solv (Solv = ROH, AN, DMF, DMSO; X = Cl, Br, I, SCN)	Ni(En)$_n$X$_2$ (n = 2, 3) Ni(En)$_2$(NH$_3$)$_2$X$_2$	164, 165
Ni–NH$_4$I–En–Solv (Solv = DMF, DMSO)	Ni(En)$_3$I$_2$·nSolv (n = 1, 2)	164, 165
PbO–NH$_4$X–En–Solv (Solv = DMF, DMSO, CH$_3$OH; X = Cl, Br, I)	Pb(En)$_n$X$_2$ (n = 1, 2) Pb(En)DMSO)I$_2$	166–169
M–3NH$_4$SCN–3En–Solv (M = Ni, Zn, Cd; Solv = CH$_3$OH, AN)	M(En)$_3$(SCN)$_2$·NH$_4$SCN	170, 171
M–2NH$_4$SCN–M^{I}SCN–3En–Solv (M = Cu, Ni, Zn, Cd; M^I = Na, K; Solv = CH$_3$OH, AN)	M(En)$_3$(SCN)$_2$·M^{I}SCN	171–173
ZnO–NH$_4$SCN–Den–Solv (Solv = CH$_3$OH, AN)	Zn(Den)$_n$(SCN)$_2$ (n = 1, 2)	174
Ni–NH$_4$SCN–Ten–AN	(HTen)$_2$Ni(SCN)$_4$	175
Ni–NH$_4$SCN–Ten–DMF	[Ni(Ten)$_2$(SCN)$_2$(H$_2$O)$_2$]·4DMF	176

Den, diethylenediamine; Ten, triethylenediamine.

$$2Ni + 6NH_4X + 2En + O_2 = Ni(NH_3)_6X_2 + Ni(En)_2X_2 \quad (3.34)$$
$$+ 2H_2O + 2HX$$
$$X = Cl, Br$$

The hexammine–nickel chloride and bromide are slightly soluble in methanol and DMF and precipitate from the reaction mixture as a blue solid. Ammonium thiocyanate in methanol and ammonium bromide in DMSO give bis(diethylenediamine)–nickel complexes. Mixed complexes of $Ni(NH_3)_2(En)_2X_2$ are obtained from ammonium iodide in methanol, and ammonium chloride in DMSO, though such compounds are more typical for $Ni/En = 1:2$ (reaction (3.35)).

$$Ni + 2NH_4X + 2En + \tfrac{1}{2}O_2 = Ni(NH_3)_2(En)_2X_2 + 2H_2O \quad (3.35)$$

When $M(MO)/En = 1:2–3$, bis- and tris(ethylenediamine) complexes are obtained, (reaction (3.36)), but iodide and thiocyanate may form solvates with DMF or DMSO. The excess of ammonium salt in the system with ethylenediamine gives double salts for thiocyanate (reaction (3.36), $M = Zn, Cd, Ni$).

$$M + \tfrac{1}{2}O_2 + 3NH_4SCN + 3En = M(En)_3(SCN)_2 \cdot NH_4SCN$$
$$+ 2NH_3 + H_2O$$

$$(3.36)$$

Sodium and potassium salts are obtained in the same manner (reaction (3.37), $M = Cu, Ni, Zn, Cd$; $M^I = Na, K$).

$$M + 2NH_4SCN + M^ISCN + 3En + \tfrac{1}{2}O_2 = M(En_3)(SCN)_2 \cdot$$
$$M^ISCN + 2NH_3 + H_2O$$

$$(3.37)$$

Double salts of copper, zinc and cadmium are formed readily in methanol when ammonium thiocyanate is completely substituted by alkaline metal salts (reaction (3.38), $M^I = Na, K$). The compounds are readily dissolved in water, DMF and DMSO. Their composition is not changed on precipitation from these solvents or from pyridine.

$$M + 3M^ISCN + 3En + 2CH_3OH + \tfrac{1}{2}O_2 = M(En)_3(SCN)_2 \cdot$$
$$M^ISCN + 2M^IOCH_3 + H_2O$$

$$(3.38)$$

Strengthening of the amine basicity in ethylenediamine derivatives leads to their protonization and the formation of anionic complexes. Thus, the complex $(HTen)_2[Ni(SCN)_4]$, in which the protonized amine is coordinated and functions as a cation, is isolated from acetonitrile when formed by reaction (3.39).

$$Ni + 4NH_4SCN + 2Ten + \tfrac{1}{2}O_2 = (HTen)_2Ni(SCN)_4$$
$$+ 4NH_3 + H_2O$$

$$(3.39)$$

The $[Ni(Ten)_2(SCN)_2(H_2O)_2]\cdot4DMF$ complex was obtained by acetonitrile substitution by DMF. Not a single solvent molecule is co-ordinated [176].

Experimental procedures

Synthesis of Cu(En)₂I₂·0.33DMF

Copper powder (0.63 g, 0.01 mol), NH_4I (2.9 g, 0.02 mol), DMF (15 cm^3) and ethylenediamine (1.34 cm^3, 0.02 mol) were placed in a flask in the above order, and the mixture was heated to 60°C and stirred until total dissolution of copper was observed (5–10 min). After cooling, blue crystals of the product precipitated from the solution. These were filtered off, washed with isopropanol and finally dried in vacuo at room temperature (yield 3.61 g; 78%) [163].

Synthesis of Pb(En)I₂·DMSO

PbO (2.23 g, 0.01 mol), NH_4I (2.90 g, 0.02 mol), DMSO (20 cm^3) and ethylene-diamine (1.34 cm^3, 0.02 mol) were placed in a flask in the above order, and the mixture was heated to 60°C and stirred until total dissolution of the lead oxide was observed (4 h). The resulting colorless solution was allowed to stand at room temperature for 24 h, during which crystals of Pb(En)I₂·DMSO precipitated. They were filtered off, washed with isopropanol and finally dried in vacuo at room temperature (yield 3.7 g; 62%) [167].

Synthesis of [Cd(En)₃](SCN)₂·KSCN

CdO (0.64 g, 0.005 mol), NH_4SCN (0.8 g, 0.01 mol), KSCN (0.49 g, 0.005 mol), CH_3CN (20 cm^3) and ethylenediamine (1 cm^3) were placed in a flask in the above order, and the mixture was heated to 60°C and stirred until total dissolution of the cadmium oxide was observed (20 min). After cooling, colorless crystals of [Cd(En)$_3$](SCN)$_2$·KSCN precipitated from the solution. These were filtered out, washed with isopropanol and finally dried in vacuo at room temperature (yield 2.32 g; 92%) [173].

Synthesis of Zn(Den)$_2$(SCN)$_2$

ZnO (0.4 g, 0.005 mol), NH$_4$SCN (0.8 g, 0.01 mol), diethylenediamine (0.86 g, 0.01 mol) and CH$_3$CN (25 cm^3) were placed in a flask and the mixture was heated to 60°C and stirred until total dissolution of the zinc oxide was observed (15 min). After cooling, colorless crystals of Zn(Den)$_2$(SCN)$_2$ precipitated from the solution. These were filtered off, washed with isopropanol and finally dried in vacuo at room temperature (yield 1.60 g; 90%) [174].

Synthesis of [Ni(SCN)$_4$(HTen)$_2$]

Nickel powder (0.58 g, 0.01 mol), NH$_4$SCN (3.04 g, 0.04 mol), triethylenediamine (2.25 g, 0.02 mol) and CH$_3$CN (25 cm^3) were placed in a flask and the mixture was heated to 60°C and stirred until total dissolution of the nickel was observed (3–4 h). Pale green crystals of [Ni(SCN)$_4$(HTen)$_2$] precipitated from the solution with 70 cm^3 of dry isopropanol. These were filtered off, washed with isopropanol and finally dried in vacuo at room temperature (yield 4.89 g; 95%) [175].

3.5.5 Complexes with aminoalcohols

Monoethanolamine and 2-dimethylaminoethanol were used in the direct synthesis as ligands and solvents. Copper and lead form similar compounds of the M(Ea)(Hea)$_n$X ($n = 0, 0.33, 1, 2$) type, containing the aminoalcohol as a coordinated anion (Ea) and a neutral ligand (HEa) (Table 3.13). The complex composition does not change with the excess of ammonium salt (for copper/NH$_4$X ratios up to 1:6).

Unlike monoethanolamine, complexes of 2-dimethylaminoethanol in the form of an anion and neutral ligand were not obtained. Compounds of Pb(Me$_2$Ea)X (Table 3.13) were obtained for lead, regardless of the solvent and the PbO/NH$_4$X ratio. Copper complexes were isolated only at the optimal ratio of Cu/NH$_4$X/HMe$_2$Ea = 1:2:2. The aminoalcohol is either an anion, or a neutral ligand, depending on the nature of the ammonium salt anion.

Diethanolamine may appear in singly (HDea) and doubly deprotonated (Dea) forms (Table 3.13). The composition of the lead complexes, obtained by the action of ammonium bromide or nitrate in DMF, depends on the ratio of the initial components. Thus, the solvate $(PbNO_3)_2(Dea)\cdot DMF$ was obtained at the ratio of $PbO/H_2Dea = 1:0.5$, and anionic complex $(H_2Dea)_2[Pb_2Br_6]$ was formed when the PbO/NH_4Br ratio was $1:3$. Triethanolamine may be mono- (H_2Tea) or di-deprotonized (HTea), or a neutral ligand (H_3Tea) in the complexes obtained by direct synthesis (Table 3.13). The composition of the copper complexes is determined by the nature of the solvent and anion, as well as by the ratio of the components. Mixed complexes with chloride, bromide and thiocyanate anions were formed at the ratio $Cu/NH_4X/H_3Tea = 1:1:1$ (reaction (3.40)).

$$Cu + NH_4X + H_3Tea + \tfrac{1}{2}O_2 = Cu(H_2Tea)(NH_3)X + H_2O$$

$$(3.40)$$

The thiocyanate complex of this composition is unstable and loses ammonia even at room temperature, changing in color from blue to green. It was isolated only from DMF. The complex with neutral triethanolamine was

TABLE 3.13
Composition of copper and lead complexes with aminoalcohols

Initial system	Compound
$Cu-NH_4X-Hea$ (X = Cl, Br, I, NO_3, Ac, SCN)	$Cu(Ea)X(HEa)_n$ ($n = 1, 2$)
$PbO-NH_4X-Hea$ (X = Cl, Br, I, NO_3)	$Pb(Ea)X(HEa)_n$ ($n = 0, 0.33, 1$)
$Cu-NH_4X-HMe_2Ea-H_2O$ (X = Cl, Br, SCN)	$Cu(Me_2Ea)X$
$Cu-NH_4X-HMe_2Ea-H_2O$ (X = NO_3, Ac)	$Cu(HMe_2Ea)X$
$PbO-NH_4X-HMe_2Ea-Solv$ (Solv = CH_3OH, AN, DMF, HMe_2Ea; X = Cl, Br, I, NO_3, SCN)	$Pb(Me_2Ea)X$
$Cu-NH_4X-H_2Dea-H_2O$ (X = Cl, Br, I, NO_3)	$Cu(HDea)X(NH_3)\cdot H_2O$
$PbO-NH_4X-H_2Dea-Solv$ (Solv = CH_3OH, DMF, DMSO; X = Cl, Br, I, NO_3, SCN)	$Pb(HDea)X$
$Cu-NH_4X-H_3Tea-Solv$ (Solv = CH_3OH, DMF; X = Cl, Br)	$Cu(H_2Tea)X(NH_3)$
$PbO-NH_4X-H_3Tea-Solv$ (Solv = CH_3OH, AN, DMF, DMSO)	$(PbX)_2(HTea)$ (X = I, Ac, SCN) $Pb(H_3Tea)X_2$ (X = Br, NO_3, Ac)

obtained with the component ratio $Cu/NH_4Br/H_3Tea = 1:2:2$ (reaction (3.41)).

$$Cu + 2NH_4Br + 2H_3Tea + \tfrac{1}{2}O_2 = Cu(H_3Tea)_2(NH_3)Br_2 \\ + NH_3 + H_2O \tag{3.41}$$

The ammine complex was obtained in aqueous and methanolic solutions of ammonium nitrate in the presence of triethanolamine, which obviously facilitates complex formation according to reaction (3.42).

$$Cu + 4NH_4NO_3 + \tfrac{1}{2}O_2 = Cu(NH_3)_4(NO_3)_2 + 2HNO_3 + H_2O \tag{3.42}$$

The composition of the lead complexes is also influenced by the ammonium salt anion and by the ratio of the components. Two types of complexes were isolated: $PbX \cdot 0.5HTea$ and $PbX_2 \cdot 2H_3Tea$. The composition of the iodide and thiocyanate complexes is not influenced by the nature of the solvent or the component ratio. Thus, $PbSCN \cdot 0.5HTea$ was obtained with a four-fold excess of triethanolamine, and the same iodide compound resulted with a three-fold excess of ammonium iodide. The composition of the nitrate and acetate complexes depends on the PbO/H_3Tea ratio (Table 3.13).

Mixed complexes were obtained with ethylenediamine and monoethanolamine. It was found that both copper and zinc form mainly two types of compounds: $M(En)(HEa)_2X_2$ and $M(En)_2(HEa)X_2$ [190,191]. Their composition depends on the quantity of ethylenediamine in the reaction mixture, except in the case of copper chloride, zinc bromide and iodide complexes. $Zn(En)_2(HEa)I_2 \cdot 0.5En$, obtained in the system with NH_4I at a ratio $ZnO/En = 1:3$, contained both coordinated and noncoordinated ethylenediamine, as shown by X-ray analysis.

Experimental procedures

Synthesis of [Cu(Ea)(HEa)₂]I

Copper powder (0.63 g, 0.01 mol), NH_4I (1.45 g, 0.01 mol) and HEa (5 cm^3) were placed in a flask and the mixture was heated to 60°C and stirred until total dissolution of the copper was observed (8–10 min). After cooling, blue crystals of the product precipitated from the solution. Those were filtered off, washed with isopropanol, and finally dried in vacuo at room temperature (yield 1.88 g; 51%) [178].

Synthesis of Pb(Me$_2$Ea)Br

PbO (2.23 g, 0.01 mol), NH$_4$Br (2.0 g, 0.02 mol), DMF (15 cm^3) and HMe$_2$Ea (10 cm^3, 0.02 mol) were placed in a flask in the above order, and the mixture was refluxed and stirred until all the solids had dissolved (3–5 min). The resulting color-less solution was allowed to stand at room temperature for 24 h, during which the crystals of Pb(Me$_2$Ea)Br separated. These were filtered off, washed with dry isopropanol and finally dried in vacuo at room temperature (yield 3.0 g; 80%) [183].

Synthesis of [Pb$_4$I$_4$(HTea)$_2$]$_n$

PbO (2.23 g, 0.01 mol), NH$_4$I (1.45 g, 0.01 mol), DMF (15 cm^3) and H$_3$Tea (1.33 cm^3, 0.01 mol) were placed in a flask in the above order, and the mixture was heated to boiling and stirred until total dissolution of the lead oxide was observed (3–5 min). The resulting colorless solution was allowed to stand at room temperature for 2 h, after which the crystals of [Pb$_4$I$_4$(HTea)$_2$]$_n$ separated from the solution. These were filtered off, washed with dry isopropanol and finally dried in vacuo at room temperature (yield 3.0 g; 74%) [181].

Synthesis of [Zn(En)$_2$(HEa)]I$_2$·0.5En

ZnO (0.81 g, 0.01 mol), NH$_4$I (2.90 g, 0.02 mol), HEa (10 cm^3) and ethylenedia-mine (2 cm^3, 0.03 mol) were placed in a flask in the above order, and the mixture was heated to boiling and stirred until total dissolution of the zinc oxide was observed (20 min). Colorless crystals of [Zn(En)$_2$(HEa)]I$_2$·0.5En precipitated from the solution with 70 cm^3 of isopropanol. These were filtered off, washed with dry isopropanol and finally dried in vacuo at room temperature (yield 4.0 g; 75%) [191].

3.6 INTERACTION OF METALS WITH QUINONES AND PHTHALOCYANINE PRECURSORS

o-Quinones are extremely reactive ligands [112,192–205], especially those containing *tert*-butyl substituents. Complexes of *o*-benzoquinones, *o*-semi-quinones and catechols of this type have been obtained by direct interaction between metal powders and the corresponding ligands. They are radical in nature and play an important role in coordination chemistry [199–201]. In nonaqueous solvents (hexane, dimethoxyethane, toluene, THF and chloro-form), the formation of complexes mainly of types **I–IV** (R = 3,5- or 3,6-*t*-Bu$_2$) has been observed by EPR.

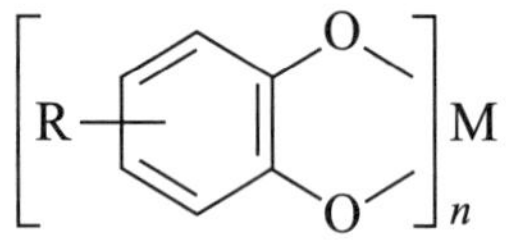

The formation of radical *o*-semiquinolates of type **I** has been detected for thallium [202], indium [195], zinc [193,196,203], cadmium [193,196,203], magnesium [196], barium [196], aluminum [193], gallium [198] and tin [192]. The adducts **II** are characteristic of copper and silver (L = PPh$_3$, $m = 2$, $n = 1$) [112] and also of indium (L = phen, $m,n = 1$) [195]. Data on the radical-anion salts of types **III** and **IV** have been presented in a review [112] and other publications [192,203]. Metal complexes of catechols can also be obtained from tetrahalo-*o*-quinones, for example, Sn(X$_4$C$_6$O$_2$)$_2$ (X = Cl, Br) [195].

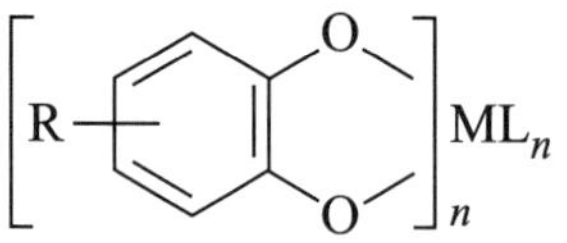

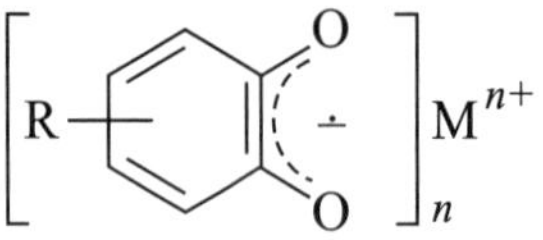

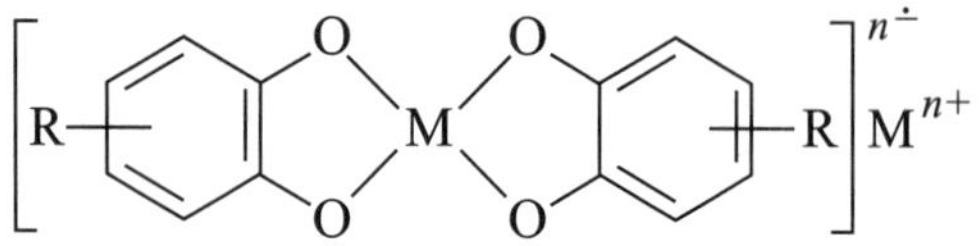

The data of Ozarovski et al. [196] and Adams et al. [198] are of the greatest interest for the study of the properties and structures of type **I** complexes. Magnesium, barium, zinc and cadmium complexes of 3,5-di-*tert*-butyl-1,2-*o*-benzoquinone (L) and their adducts with pyridine, 2,2′-bipyridyl, and *N,N,N′,N′*-tetramethylethylenediamine (L′), having the compositions ML$_2$ and ML$_2$·nL′($n = 1, 2$), were isolated preparatively and characterized in the first of these studies. Complexes **I** and their adducts with L′ have been obtained by the reaction of the above elemental metals and ligands in toluene; detailed EPR spectroscopic studies confirmed their biradical nature and made it possible to establish the conformations of the biradical ligands. The study of Adams et al. [198] was devoted to the triradical gallium complex of 3,5-di-*tert*-butyl-1,2-*o*-benzoquinone **I** ($n = 3$). This compound was obtained by heating gallium and the ligand in boiling toluene (under an argon atmosphere) and was characterized by X-ray diffraction. The complex has the form of a three-bladed propeller with the gallium atom in the center: the length of the C–C bond between the chelating oxygen atoms is 1.439(12) Å, which indicates the semiquinone nature of the ligands. It is striking that the above value differs little from the same bond (1.433 Å) in the chromium analogue, the structure of which had been characterized previously [205]. The gallium complex **I** ($n = 3$) exhibits ferromagnetic properties; its magnetic moment is 2.95 B.M. at 320 K and 3.58 B.M. at 9 K; it then decreases again and is 3.26 B.M. at 2 K. Some products of the interaction of quinones and elemental metals are presented in Table 3.14.

Thus direct evidence for the formation of radical complexes in the reactions of elemental metals with o-quinones was obtained in the studies quoted above, methods were developed for the synthesis of these coordination compounds, and the characteristics of their structure and their physicochemical properties were established.

In a number of instances, the interaction of elemental metals with the ligand is accompanied by a template reaction, which yields macrocyclic chelates. The template transformation involving the direct synthesis of the macrocyclic phthalocyanine (Pc) complexes **V** from transition metals and o-phthalodinitrile (1,2-cyanobenzene), which has long been known, has now become a classical reaction [210–212].

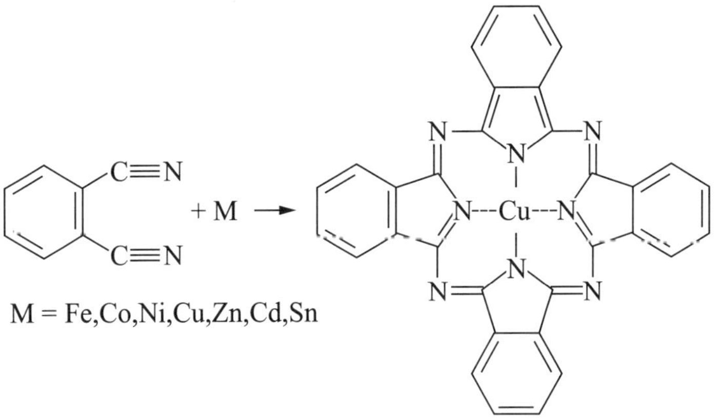

M = Fe,Co,Ni,Cu,Zn,Cd,Sn

V

The phthalocyanine complexes **V** are formed in high-boiling solvents: α-methyl- and α-chloronaphthalene, o-dichloronitro- and trichlorobenzene, and quinoline. Influence of a solvent on metal-free and metal phthalocyanines formation is described in detail in [199] of Chapter 2. Apart from phthalonitrile, o-cyanobenzamide, phthalimide and 1,3-diiminoizoindoline have been introduced into this reaction [210,211] (see also the additional data in Chapter 2 on the electrosynthesis of the phthalocyanines). A series

TABLE 3.14
Reaction products of metals and quinones

Metal	Quinone + substrate	Product	Ref.
Ag, Cu, Hg	(1) 3,5-Di-*t*-butyl-1,2-benzoquinone (2) 3,6-Di-*t*-butyl-1,2-benzoquinone + dimethoxyethane (3) Phenanthrenequinone + dimethoxyethane (4) *o*-Chloranil	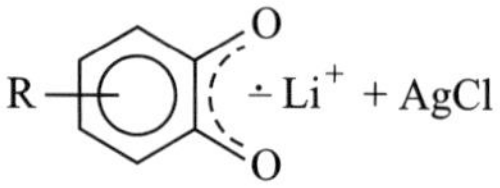 R—[ ] $\dot{=}$ Li$^+$ + AgCl	206, 207
In	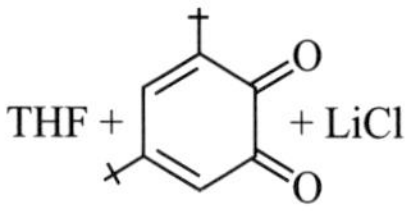 THF + [ ] + LiCl	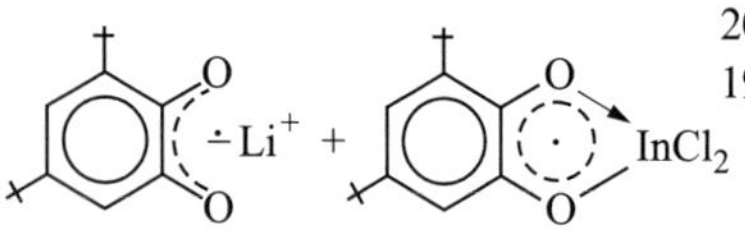 [ ] $\dot{=}$ Li$^+$ + [ ]InCl$_2$ In presence of *phen* or Et$_2$O: In L·*phen* or InL·Et$_2$O	206 194
Cu, Hg	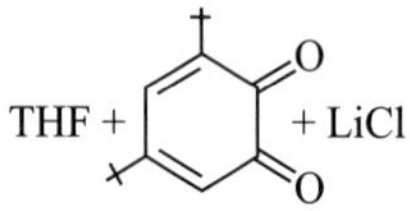 THF + [ ] + LiCl	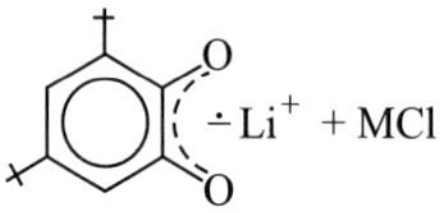 [ ] $\dot{=}$ Li$^+$ + MCl	206
Cu	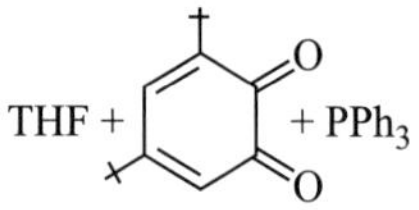 THF + [ ] + PPh$_3$	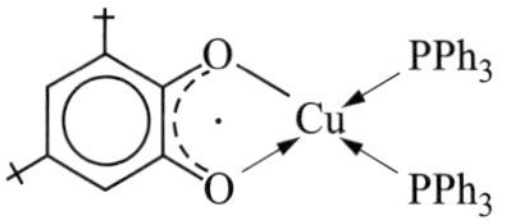 [ ]Cu(PPh$_3$)(PPh$_3$)	207
Cu, Ag	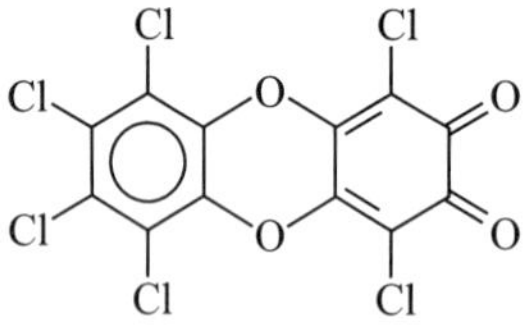 [ ]	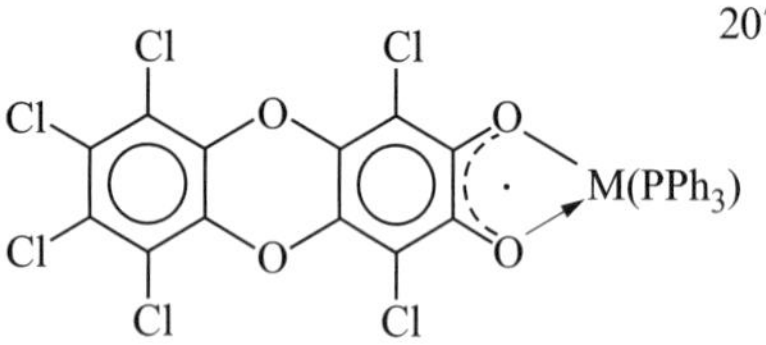 [ ]M(PPh$_3$)	207
Tl	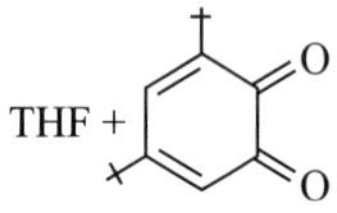 THF + [ ]	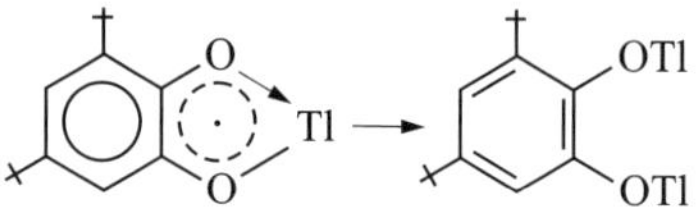 [ ]Tl → [ ]OTl, OTl	202
Tl	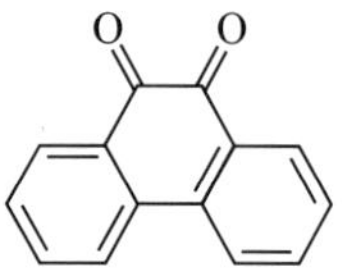 [ ] = L + THF	TlL	202

TABLE 3.14   (Contd.)

Metal	Quinone + substrate	Product	Ref.
Amalgam of Zn or Cd	THF + 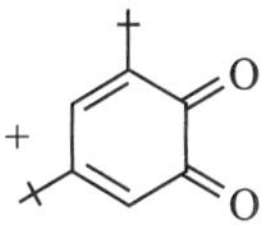	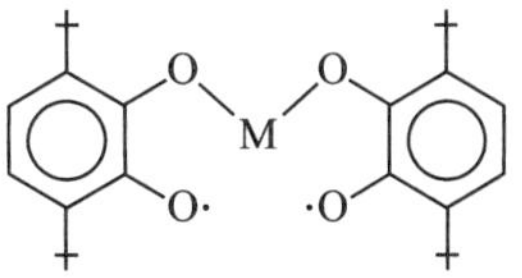	204
Ga	Tetrahalogeno-o-benzoquinones $X_4C_6O_2$, toluene, N_2 (X = Cl, Br) In the presence of I_2 and phen	$Sn^{IV}(X_4C_6O_2)$ $SnI_2(X_4C_6O_2)\cdot phen$	195
Sb	Tetrahalogeno-o-benzoquinones $X_4C_6O_2$, Et_2O (X = Cl, Br)	$Sb^V(X_4C_6O_2)_{2.5}\cdot Et_2O$ X = Cl, $n = 1.5$; X = Br, $n = 1$	208
Te	RO_2: R = Cl_4C_6, Br_4C_6, 3,5-t-$Bu_2H_2C_6$	$Te(O_2R)_2$	

of articles has been devoted to the preparation of Pc from metal alloys ([213] and references therein). The most important advantage of using an alloy is the easier reaction between phthalonitrile and the alloy's component(s), due to a concentration gradient of metal particles on the alloy surface. Due to such an interaction, it is possible to obtain polynuclear phthalocyanines and separate the alloy [212]. This is a relatively new and intriguing subarea in Pc research and undoubtedly should be taken into account.

Experimental procedures

Synthesis of copper bis(cyclohexyl isocyanide)-3,5-di-t-butyl-1,2-benzosemiquinolate

Metallic copper (1.28 g, 20 mmol) was added to a solution containing 3,5-di-t-butyl-1,2-benzoquinone (4.4 g, 20 mmol) and cyclohexyl isocyanide (4.5 cm^3, 50 mmol) in toluene (50 cm^3). The solution was stirred for several hours at 20°C and became red–brown. Then a red–brown solid was formed. The solution was evaporated in vacuo at 20°C to 25 cm^3 and was frozen at 0°C. The solid was filtered and dried in vacuo for 2 h at 20°C. Yield $\approx$ 100%; m.p. 70°C (with decomposition). Analogous derivatives of other o-quinones were prepared similarly [214].

Synthesis of copper phthalocyanine from urea and phthalic anhydride

This method is widely used in industry: urea (27 g), N,N,N',N'-tetramethylurea (0.8 g), metallic copper powder (2.13 g), phthalic anhydride (19.95 g), molybdic anhydride (0.105 g) and technical grade trichlorobenzene (mixture of isomers) (72 g) were introduced in the above order into a 250 cm^3 flask fitted with an agitator, thermometer, cooler and external heating bath. The reaction mixture was heated from 20 to 170°C in 90 min with stirring and was then kept at 170°C for 5 h. The volatile portion was distilled off under vacuum (50 mmHg) by keeping the temperature of the bath at about 150°C. The residue was digested at 90°C for 2 h, while being stirred, with 560 g of an aqueous solution of H_2SO_4 (10% w/w). The filtrate, after drying in an oven at 100°C for 24 h, comprised 17.3 g of copper phthalocyanine in the β-crystalline form. Yield 89.6% [215].

Synthesis of copper phthalocyanine from phthalonitrile

    In Linstead's classic experiments, a mixture of phthalonitrile (4 mmol) and copper (1 mmol) was heated (by oil-bath) with stirring in a wide glass tube. A green color first formed at 190°C, then the mass became pasty at 220°C and was too stiff to be stirred at 270°C (10 min). At a bath temperature of about 220°C the internal temperature began to mount rapidly and at times exceeded that of the bath by 45°C. The mass was left for another 5 min in the bath, cooled slightly and ground with alcohol. The finely powdered product was boiled repeatedly with alcohol until the washings were colorless and contained no phthalonitrile; it was then dried. Yield 75–90% of the weight of phthalonitrile [216]. In more recent publications, CH_3ONa and other strong bases are used in order to perform a nucleophilic attack at the cyano group of phthalonitrile (scheme (3.43)) [217,218].

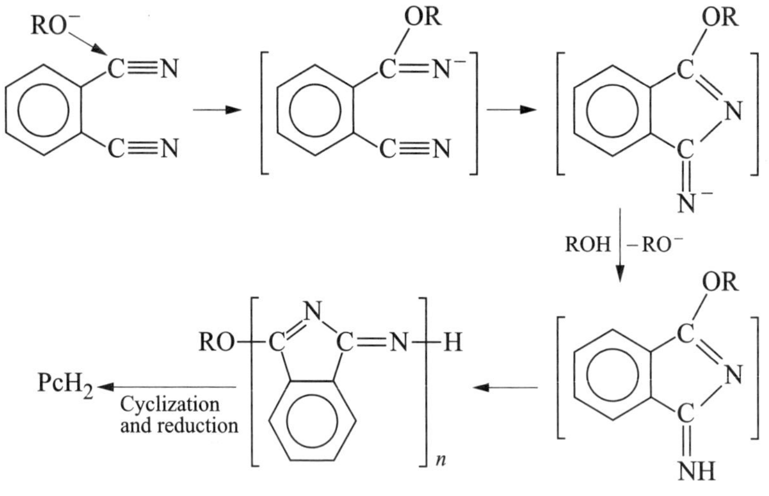

$$(3.43)$$

Synthesis of magnesium phthalocyanine from o-cyanobenzamide

Linstead performed classic experiments on this method: pure *o*-cyanobenzamide (10 g) was heated for 15 min at 230–240°C with magnesium metal (2 g). The melt was powdered, extracted successively with dilute NaOH solution, 10% sulfuric acid (overnight) and hot water. After prolonged extraction with boiling alcohol, bright blue solid remained [219]. Yield 4 g.

Metal alloys + phthalocyanine precursor

The alloys were mixed with phthalonitrile and pressed into pellets. The pellets were inserted into a glass ampoule, evacuated and sealed off. The ampoule was heated at about 480 K for several days. Reaction products (SnPc from Au–Sn alloy, PbPc from Au–Pb alloy, In_2Pc_3 and SnPc from Sn–In alloy, In_2Pc_3 and Tl_2Pc from In–Tl alloy, etc.) were identified by X-ray diffraction methods on single crystals and/or powdered samples [213].

REFERENCES

1. Kakovsky, N.A.; Potashnikov, Yu.M. *Kinetics of Dissolution Processes.* Moscow: Metallurgia (1975).
2. Levich, V.G. *Physico-Chemical Hydrodynamics.* Moscow: Gosizdat.Fiz.-Mat.Lit. (1959).
3. Pleskov, Yu.V.; Filinovsky, V.Yu. *Rotating Disc Electrodes.* Moscow: Nauka (1972).
4. Dolivo-Dobrovolsky, V.V. *Zapiski Leningradskogo Gornogo Inst.* **42**(3), 3 (1963).
5. Dolivo-Dobrovolsky, V.V. *Obogashenie Rud.* **53**(5), 38 (1964).
6. Dolivo-Dobrovolsky, V.V. *Zapiski Leningradskogo Gornogo Inst.* **50**(3), 43 (1970).
7. Zelikman, A.N.; Voldman, G.M.; Belyaevskaya L.V. *Theory of Hydrometallurgy Processes.* Moscow: Metallurgy (1983).
8. Maslenitzky, I.N.; Chugaev, L.V. *Metallurgy of Noble Metals.* Moscow: Metallurgy (1972).
9. Zavgoronya, E.F.; Lubyanova, V.I.; Degtyareva, Z.E. *Zh. Prikl. Khim. (Leningrad).* **52**(11), 2583 (1979).
10. Rozovsky, A.Ya. *Kinetics of Topochemical Reactions.* Moscow: Khimiya (1974).
11. Makhmudova, E.A.; Nurullaev, S.P., Rustamov, C.R. *Dokl. Akad. Nauk Uzbek, SSR.* No. 5, 33 (1988).
12. Ospanov, C.K.; Aytchozhaeva, E.A. *Khimia i Khimicheskaya Tekhnologia.* **20**, 172 (1976).
13. Delmon, B. *Kinetics of Heterogeneous Reactions.* Moscow: Mir (1972).
14. Sharov, S.K.; Ivanov, L.P.; Gorbunov, A.I. *Metal-Organic Components of Catalysts.* Moscow: Nauka (1986).
15. Du Preez, J.G.H; Morris, D.C.; Van Vuuren, C.P.J. *Hydrometallurgy* **6**(3–4), 197 (1981).
16. Zenchenko, D.A.; Gorichev, I.G.; Zenchenko, A.D. *Koord. Khim.* **15**(7), 934 (1989).
17. Gorichev, I.G.; Zenchenko, D.A.; Michalchenko, I.S.; Serochov, V.D. *Koord. Khim.* **12**(7), 886 (1986).
18. Sidhu, P.S.; Gilkes, R.J.; Cornell, R.M. et al. *Clays Clay Miner.* **29**(4), 269 (1981).

19. Gorichev, I.G.; Kiprianov, N.A. *Zh. Prikl. Khim. (Leningrad).* **50**(3), 503 (1977).

20. Gorichev, I.G.; Gorsheneva, V.F.; Kiprianov, N.A.; Klyuchnikov, N.G. *Kinet. Katal.* **21**(6), 1422 (1980).

21. Ashharua, F.G.; Gorichev, I.G.; Klyuchnikov, N.G. *Zh. Prikl. Khim. (Leningrad)* **49**(2), 318 (1976).

22. Gorichev, I.G.; Ashharua, F.G.; Vainman, S.K.; Klyuchnikov, N.G. *Zh. Fiz. Khim.* **50**(6), 1610 (1976).

23. Machmudova, E.A.; Nurullaev, S.P.; Karimov, C.S.; Rustamov, C.R. *Uzbek. Khim. Zh.* **6**, 3 (1988).

24. Gorichev, I.G.; Malov, L.V.; Duchanin, V.S. *Zh. Fiz. Khim.* **52**(5), 1195 (1978).

25. Gorichev, I.G.; Shevelev, N.P.; Malov, L.V.; Duchanin, V.S. *Zh. Fiz. Khim.* **56**(5), 1154 (1982).

26. Gorichev, I.G.; Kiprianov, N.A. *Uspekhi Khim.* **53**(11), 1790 (1984).

27. Bessarabov, A.M.; Kushnir, E.Yu.; Allahberdov, G.R. *Zh. Fiz. Khim.* **58**(10), 2588 (1984).

28. Ginstling, A.M.; Brounstein, B.I. *Zh. Prikl. Khim. (Leningrad)* **23**(12), 1250 (1950).

29. Prafulla, R.R.; Sharad, G. D. *J. Chem. Tech. Biotechnol.* **42**, 167 (1988).

30. Rath, P.C.; Paramguru, R.K.; Jena, P.K. *Hydrometallurgy* **6**, 219 (1981).

31. Redin, V.I.; Novikova, N.M.; Novikov, Yu.A.; Spielfogel, P.V. Deposited In Niiitekhim. Cherkassy, 331hp–D82 (1982).

32. Pchelnikova, R.I.; Batrakov, V.V.; Gorichev, I.G.; Ryagusov, A.I. *Zh. Prikl. Khim.* **63**(2), 429 (1990).

33. Gorichev, I.G.; Michalchenko, I.S. *Koord. Khim.* 1986, **12**(8), 1082 (1986).

34. Prodan, E.A. *Inorganic Topochemistry*. Minsk: Nauka y Technika (1986).

35. Gorichev, I.G.; Klyuchnikov, N.G.; Bibikova, Z.N.; Popova, L.F. *Zh. Fiz. Khim.* **50**(5), 1189 (1976).

36. Shevelev, N.P.; Gorichev, I.G.; Klyuchnikov, N.G. *Zh. Fiz. Khim.* **48**(11), 2750 (1974).

37. Erofeev, B.V.; Protashchik, V.A. *Dokl. Akad. Nauk USSR.* 155(3), 647 (1964).

38. Gorbachev, V.M. *Zh. Fiz. Khim.* **49**(10), 2415 (1975).

39. Erofeev, B.V. *Dokl. Akad. Nauk USSR.* **52**(6), 515 (1946).

40. Erofeev, B.V. Izv. *Akad. Nauk Byelorus, SSR* No. 4, 137 (1950).

41. Sakovich, G.V. *Uch. Zapiski Tomskogo Gos. Univ.* No. 26, 103 (1955).

42. Pavlenko, V.A.; Skopenko, V.V.; Kokozay, V.N. *Dokl. Akad. Nauk Ukr. SSR, Ser. B* No. 7, 47 (1983).

43. Skopenko, V.V.; Kokozay, V.N.; Nevesenko, N.D. *Dokl. Akad. Nauk Ukr. SSR, Ser. B* (4), 3 (1984).

44. Pavlenko, V.A.; Kokozay, V.N.; Skopenko, V.V. *Ukr. Khim. Zh.* **51**(4), 342 (1985).

45. Pavlenko, V.A.; Kokozay, V.N.; Skopenko, V.V. *Dokl. Akad. Nauk Ukr. SSR, Ser. B* (9), 46 (1985).

46. Pavlenko, V.A.; Kokozay, V.N. *Ukr. Khim. Zh.* **52**(7), 700 (1986).

47. Nevesenko, N.D.; Kokozay, V.N. *Ukr. Khim. Zh.* **52**(11), 1129 (1986).

48. Nevesenko, N.D.; Skopenko, V.V.; Kokozay, V.N. *Zh. Neorg. Khim.* **31**(8), 2062 (1986).

49. Skopenko, V.V.; Kokozay, V.N.; Nevesenko, N.D. *Dokl. Akad. Nauk Ukr. SSR, Ser. B* (10), 47 (1986).

50. Nevesenko, N.D.; Kokozay, V.N.; Shvachko, V.I. *Dokl. Akad. Nauk Ukr. SSR, Ser. B* (9), 51 (1987).

51. Vassilyeva, O.Yu.; Nevesenko, N.D.; Kokozay, V.N. et al. *Dokl. Akad. Nauk Ukr. SSR, Ser. B* (3), 36 (1988).

52. Dudarenko, N.M.; Kokozay, V.N. *Ukr. Khim. Zh.* **55**(2), 129 (1989).

53. Pavlenko, V.A.; Kokozay, V.N.; Skopenko, V.V. *Dokl. Akad. Nauk Ukr. SSR, Ser. B* (6), 48 (1989).

54. Kokozay, V.N.; Polyakov, V.R.; Skopenko, V.V. *Dokl. Akad. Nauk Ukr. SSR, Ser. B* (11), 47 (1989).

55. Pavlenko, V.A.; Kokozay, V.N. *Zh. Prikl. Khim.* **63**(1), 216 (1990).

56. Dudarenko, N.M.; Kokozay, V.N. *Izv. VUZov. Ser. Khim. Khim. Tekhnol.* **33**(3), 12 (1990).

57. Polyakov, V.R.; Kokozay, V.N. *Dokl. Akad. Nauk Ukr. SSR, Ser. B* (7), 58 (1990).

58. Polyakov, V.R.; Kokozay, V.N. *Ukr. Khim. Zh.* **56**(9), 912 (1990).

59. Polyakov, V.R., Kokozay, V.N. *Dokl. Akad. Nauk Ukr. SSR, Ser. B* (10), 51 (1990).

60. Polyakov, V.R.; Kokozay, V.N.; Pavlenko, V.A. *Ukr. Khim. Zh.* **57**(2), 127 (1991).

61. Vassilyeva, O.Yu.; Kokozay, V.N. *Dokl. Akad. Nauk Ukr. SSR, Ser. B* (9), 119 (1992).

62. Dudarenko, N.M.; Kokozay, V.N.; Pavlenko, V.A. *Ukr. Khim. Zh.* **58**(9), 721 (1992).

63. Dudarenko, N.M.; Kokozay, V.N.; Pavlenko, V.A. *Ukr. Khim. Zh.* **58**(10), 859 (1992).

64. Vassilyeva, O.Yu.; Kokozay, V.N. *Ukr. Khim. Zh.* **59**(2), 176 (1993).

65. Vassilyeva, O.Yu.; Kokozay, V.N. *Ukr. Khim. Zh.* **59**(12), 1249 (1993).

66. Vassilyeva, O.Yu.; Kokozay, V.N. *Dokl. Akad. Nauk Ukr. SSR, Ser. B* (9), 136 (1993).

67. Kokozay, V.N.; Nevesenko, N.D. *XI Ukrainian Republican Conference on Inorganic Chemistry*. Proc. Reports Uzhgorod, p. 104 (1986).

68. Lavrentiev, I.P.; Korableva, L.G.; Lavrentieva, E.L. et al. *Izv. Akad. Nauk USSR, Ser. Khim.* No. 9, 1961 (1974).

69. USSR Patent 414193; Lavrentiev, I.P.; Korableva, L.G.; Khidekel, M.L. *Bull. Izobr.* No. 5 (1974)/.

70. USSR Patent 540822; Korableva, L.G.; Lavrentiev, I.P.; Khidekel, M.L. et al. *Bull. Izobr.* No. 48 (1976).

71. Kurskov, S.N.; Ivleva, I.N.; Lavrentiev, I.P.; Filipenko, O.S.; Khidekel, M.L. *Izv. Akad. Nauk USSR, Ser. Khim.* No.7 , 1442 (1976).

72. Kurskov, S.N.; Ivleva, I.N.; Lavrentiev, I.P.; Khidekel, M.L. *Izv. Akad. Nauk USSR, Ser. Khim.* No. 8, 1708 (1977).

73. Zhukov, S.A.; Lavrentiev, I.P.; Nifontova, G.A. *React. Kinet. Catal. Lett.* **7**(4), 405 (1977).

74. Letuchii, Ya.A.; Lavrentiev, I.P.; Khidekel, M.L. *Proc. XIII Chugaev Conference on Coordination Chemistry*. Moscow, p. 234 (1978).

75. Kurskov, S.N.; Lavrentiev, I.P.; Khidekel, M.L. *Izv. Akad. Nauk USSR, Ser. Khim.* No. 4, 713 (1979).

76. Letuchii, Ya.A.; Lavrentiev, I.P.; Khidekel, M.L. *Izv. Akad. Nauk USSR, Ser. Khim.* No. 7, 718 (1976).

77. Letuchii, Ya.A.; Lavrentiev, I.P.; Khidekel, M.L. *Izv. Akad. Nauk USSR, Ser. Khim.* No. 4, 888 (1979).

78. Lavrentiev, I.P.; Korableva, L.G.; Lavrentieva, E.L.; Nikiforova, G.A. et al. *Koord. Khim.* **5**(10), 1484 (1979).

79. Lavrentiev, I.P.; Khidekel, M.L. *Proc. XIV Chugaev Conference on Coordination Chemistry*, Vol. 1, pp. 19–20 (1981).

80. Nifontova, G.A.; Lavrentiev, I.P.; Ponomarev, V.I.; Fillipov, O.S. *Izv. Akad. Nauk USSR, Ser. Khim.* No. 8, 1691 (1982).

81. Nifontova, G.A.; Lavrentiev, I.P.; Letuchii, Ya.A. et al. *Izv. Akad. Nauk USSR, Ser. Khim.* No. 9, 129 (1982).

82. Letuchii, Ya.A.; Lavrentiev, I.P.; Khidekel, M.L. *Koord. Khim.* **8**(11), 1477 (1982).

83. Grigoryan, E.A.; Konstantinov, I.E.; Lavrentiev, I.P.; Erofeeva, K.A. et al. *Zh. Fiz. Khim.* **57**(4), 929 (1983).

84. Chulkevich, A.K.; Lavrentiev, I.P.; Khidekel, M.L. *Proc. Conference "Complexes with Charge Transfer and Ion-Radical Salts"*, Chernogolovka, p. 250 (1984).

85. Kurskov, S.N.; Lavrentiev, I.P.; Khidekel, M.L. *Proc.1st Conference on Chemicals*, Ufa, p. 84 (1985).

86. Chulkevich, A.K.; Korableva, L.G.; Lavrentiev, I.P. et al. *Proc. V Conference on Non-Aqueous Solutions*, Rostov-on-Don, p. 46 (1985).

87. Lavrentiev, I.P.; Khidekel, M.L. *Proc. V Conference on Non-Aqueous Solutions*, Rostov-on-Don, p. 47 (1985).

88. Chulkevich, A.K.; Korableva, L.G.; Lavrentiev, I.P. et al. *Izv. Akad. Nauk USSR, Ser. Khim.* No. 3, 6671 (1986).

89. Chulkevich, A.K.; Lavrentiev, I.P.; Moravsky, A.P. et al. *Koord. Khim.* **12**(4), 470 (1986).

90. Chulkevich, A.K.; Lavrentiev, I.P. *Izv. Akad. Nauk USSR, Ser. Khim.* (9), 2130 (1986).

91. Nifontova, G.A.; Kaplunov, M.G.; Lavrentiev, I.P. *Proc. II Conference "Use of Vibrational Spectra in the Study of Inorganic and Co-ordination Compounds"*, Krasnoyarsk, p. 141 (1987).

92. Nifontova, G.A.; Kaplunov, M.G.; Lavrentiev, I.P. *Proc. VI Conference on Non-Aqueous Solutions*, Rostov-on-Don, pp. 143–144 (1987).

93. Shirshova, L.V.; Korableva, L.G.; Astakhova, A.S. et al. *Proc. VI Conference on Non-Aqueous Solutions*, Rostov-on-Don, p. 145 (1987).

94. Nifontova, G.A.; Krasochka, O.N.; Lavrentiev, I.P. et al. *Izv. Akad. Nauk USSR, Ser. Khim.* (2), 450 (1988).

95. Chulkevich, A.K.; Lavrentiev, I.P. *Koord. Khim.* **14**(10), 1434 (1988).

96. Lavrentiev, I.P. *Proc. XIV Chernyaev Conference on Platinum Metals*, Novosibirsk, Vol. 1, p. 26 (1989).

97. Shirshova, L.V.; Korableva, L.G.; Lavrentiev, I.P. et al. *Proc. Conference "Diffraction Methods in Chemistry"*, Suzdal, Part 2, p. 202 (1988).

98. Nifontova, G.A.; Kaplunov, M.G.; Lavrentiev, I.P. *Koord. Khim.* **15**(1), 32 (1989).

99. Shirshova, L.V.; Korableva, L.G.; Astachova, A.S.; Lavrentiev, I.P.; Ponomarev, V.I. *Koord. Khim.* **16**(3), 348 (1990).

100. Shirshova, L.V.; Ponomarev, V.I.; Korableva, L.G.; Lavrentiev, I.P. *Koord. Khim.* **16**(8), 1118 (1990).

101. Nifontova, G.A.; Filipenko, O.S.; Astachova, A.S.; Lavrentiev, I.P.; Atovmyan, L.O. *Koord. Khim.* **16**(2), 218 (1990).

102. Epelbaum, E.T.; Ponomarev, V.I.; Lavrentiev, I.P. *Izv. Akad. Nauk USSR, Ser. Khim.* No. 6, 1325 (1991).

103. Khentov, V.Ya.; Velikanova, L.N.; Lavrentiev, I.P. *Zh. Fiz. Khim.* **65**(7), 1986 (1991).

104. Nifontova, G.A.; Lavrentiev, I.P. *Izv. Akad. Nauk USSR, Ser. Khim.* No. 3, 498 (1992).

105. Shirshova, L.V.; Lavrentiev, I.P. *Izv. Akad. Nauk USSR, Ser. Khim.* (8), 192 (1992).

106. Kurskov, S.N. *Proc. III Regional Conference of Middle Asia Republics and Kazakhstan on Chemicals,* Tashkent, Vol. 2, 34 (1990).

107. Lavrentieva, E.L.; Ponomarev, V.I.; Lavrentiev, I.P. *Izv. Akad. Nauk USSR, Ser. Khim.* No. 12, 2794 (1990).

108. Lavrentiev, I.P.; Lavrentieva, E.L.; Khidekel, M.L.; Ponomarev, V.I.; Atovmyan, L.O. *Izv. Akad. Nauk USSR, Ser. Khim.* (7), 1639 (1975).

109. USSR Patent 421354; Lavrentiev, I.P.; Korableva, L.G.; Khidekel, M.L. *Bull. Izobr.* No. 12, p. 11 (1974).

110. Ryabinin, V.A.; Kondin, A.V.; Maslennikov, V.P. et al. *Zh. Obshch. Khim.* **57**(12), 2649 (1987).

111. Lavrentiev, I.P.; Khidekel, M.L. *Uspekhi Khim.* **52**(4), 569 (1983).

112. Garnovskii, A.D.; Ryabuchin, Yu.I.; Kuzharov, A.S. *Koord. Khim.* **10**(8), 1011 (1984).

113. Nifontova, G.A.; Lavrentiev, I.P. *Transition Met. Chem.* **18**, 27 (1993).

114. Shirshova, L.V.; Lavrentiev, I.P.; Ponomarev, V.I. *Koord. Khim.* **15**(8), 1048 (1989).

115. Kokozay, V.N.; Chernova, A.S.; Pavlenko, V.A. *Zh. Prikl. Khim. (Leningrad)* No. 10, 2219 (1982).

116. Skopenko, V.V.; Kokozay, V.N. *Dokl. Akad. Nauk Ukr. SSR, Ser. B* No. 9, 740 (1979).

117. Kokozay, V.N.; Pavlenko, V.A. *Ukr. Khim. Zh.* **44**(9), 921 (1978).

118. Kokozay, V.N.; Pisetzkaya, L.V.; Ulko, N.V. *Ukr. Khim. Zh.* **46**(11), 1157 (1980).

119. Xue, Gi; Jiang, S.; Xuang, X.; Shi, G.; Sun, B. *J. Chem. Soc., Dalton. Trans.* No. 6, 1487 (1988).

120. Kuzharov, A.S.; Garnovskii, A.D. *Proc. XIV Chugaev Conference on Coordination Chemistry,* Ivanovo, Part 2, p. 412 (1981).

121. Skopenko, V.V.; Kokozay, V.N.; Nevesenko, N.D. *Dokl. Akad. Nauk Ukr. SSR, Ser. B* No. 9, 42 (1980).

122. Kurskov, S.N. *Koord. Khim.* **13**(8), 1082 (1987).

123. Nifontova, G.A.; Krasochka, O.N.; Lavrentiev, I.P.; Makitov, D.D.; Atovmyan, L.O.; Khidekel, M.L. *Izv. Akad. Nauk USSR, Ser. Khim.* No. 2, 450 (1988).

124. Shirshova, L.V.; Lavrentiev, I.P.; Khidekel, M.L.; Krasochka, O.N.; Ponomarev, V.I.; *Izv. Akad. Nauk USSR, Ser. Khim.* No. 8, 1889 (1989).

125. Charles, R.G.; Barnartt, S. *J. Phys. Chem.* **62**(3), 315 (1958).

126. Sato, K.; Kammori, O. *Bull. Chem. Soc. Jpn.* **42**(10), 2790 (1969).

127. Ross, W.D.; Sievers, R.E. *Anal. Chem.* **41**(8), 1109 (1969).

128. Kokozay, V.N. *Ukr. Khim. Zh.* **45**(2), 113 (1979).

129. Skopenko, V.V.; Kokozay, V.N. *Ukr. Khim. Zh.* **46**(1), 3 (1980).

130. Kuzharov, A.S.; Garnovskii, A.D. *Proc. IV Conference "Synthesis and Study of Inorganic Compounds in Non-Aqueous Media",* Ivanovo, p. 82 (1980).

131. Barnartt, S.; Charles, R.G.; Littan, L.W. *J. Phys. Chem.* **62**(6), 763 (1958).

132. Kurskov, S.N.; Varlamov, N.V.; Labunskaya, V.I.; Koblova O.E.; Bolshakov, A.F. *Koord. Khim.* **20**(1), 70 (1994).
133. Joshi, K.C.; Pathak, V.N. *Coord. Chem. Rev.* **22**(1–2), 37 (1977).
134. Ryabuchin, Yu.I.; Kuzharov, A.S.; Garnovskii, A.D. *Proc. XV Chugaev Conference on Coordination Chemistry*, Kiev, Part 1, p. 8 (1985).
135. USSR Patent 882916; *Bull. Izobr.* No. 43, 85 (1981).
136. USSR Patent 802186; *Bull. Izobr.* No. 5, 73 (1981).
137. Polonskii, A.V.; Baryshnikov, I.A.; Achelik, V.R. *Zh. Neorg. Khim.* **27**, 531 (1982).
138. USSR Patent 1204565; *Bull. Izobr.* No. 2, 93 (1986).
139. USSR Patent 606299; *Bull. Izobr.* No. 1, 271 (1987).
140. Smirnov, V.A.; Kopetzky, I.V.; Redkin, A.N. *Vysokochist. Veshch.* No. 1, 42 (1988).
141. Smirnov, V.A.; Redkin, A.N.; Dmitriev, V.S.; Makovey, Z.I. *Vysokochist. Veshch.* No. 5, 82 (1988).
142. USSR Patent 1038283; *Bull. Izobr.* No. 32, 79 (1983).
143. Polyakov, V.R.; Kokozay, V.N.; Skopenko, V.V. *Dokl. Akad. Nauk Ukr. SSR, Ser. B* No. 11, 44 (1989).
144. USSR Patent 1710509; *Bull. Izobr.* No. 5, 86 (1992).
145. USSR Patent 1710511; *Bull. Izobr.* No. 5, 86 (1992).
146. USSR Patent 1710512; *Bull. Izobr.* No. 5, 86 (1992).
147. USSR Patent 1735269; *Bull. Izobr.* No. 19, 36 (1992).
148. USSR Patent 1038284; *Bull. Izobr.* No. 32, 79 (1983).
149. Kokozay, V.N.; Pavlenko, V.A.; Nevesenko, N.D. *Proc. V Conference on Non-Aqueous Solutions*, Moscow, p. 48 (1985).
150. USSR Patent 1186571; *Bull. Izobr.* No. 39, 99 (1985).
151. USSR Patent 1263630; *Bull. Izobr.* No. 38, 90 (1986).
152. Kokozay, V.N.; Dudarenko, N.M. *Ukr. Khim. Zh.* **53**, 1014 (1987).
153. USSR Patent 1181998; *Bull. Izobr.* No. 36, 92 (1985).
154. USSR Patent 1263631; *Bull. Izobr.* No. 38, 90 (1986).
155. USSR Patent 1650594; *Bull. Izobr.* No. 88 (1991).
156. Skopenko, V.V.; Kokozay, V.N.; Nevesenko, N.D. *Proc. IV Conference on the Chemistry of Mn, Co, and Ni*, Tbilisi, p. 167 (1983).
157. Kokozay, V.N.; Pavlenko, V.A.; Dudarenko, N.M. *Proc. XI Ukrainian Conference on Inorganic Chemistry*, Kiev, p. 105 (1986).
158. USSR Patent 1386568; *Bull. Izobr.* No. 13, 108 (1988).
159. Pavlenko, V.A.; Kokozay, V.N.; Skopenko, V.V. *Dokl. Akad. Nauk Ukr. SSR, Ser. B* No. 7, 42 (1985).
160. Kokozay, V.N.; Pavlenko, V.A.; Nevesenko, N.D. *Proc. Ukrainian Conference "Problems of Chemical Science and Industry"*, Kiev, p. 33 (1984).
161. Kokozay, V.N.; Nevesenko, N.D.; Skopenko, V.V. *Dokl. Akad. Nauk Ukr. SSR, Ser. B* **2**, 40 (1987).
162. Vassilyeva, O.Yu.; Kokozay, V.N. *Koord. Khim.* **17**(7), 968 (1991).
163. Vassilyeva, O.Yu.; Kokozay, V.N.; Simonov, Yu.A. *Zh. Neorg. Khim.* **36**(12), 3119 (1991).

164. Kokozay, V.N.; Pavlenko, V.A. *Proc. Ukrainian Conference on Inorganic Chemistry,* Simferopol, p. 145 (1981).

165. USSR Patent 1175873; *Bull. Izobr.* No. 32, 104 (1985).

166. Kokozay, V.N.; Polyakov, V.R.; Dvorkin, A.A. *Zh. Neorg. Khim.* **37**(1), 64 (1992).

167. Kokozay, V.N.; Polyakov, V.R.; Simonov, Yu.A. *J. Coord. Chem.* **28**, 191 (1993).

168. Polyakov, V.R.; Kokozay, V.N. *J. Coord. Chem.* **32**, 343 (1994).

169. Kokozay, V.N.; Sienkiewicz, A.V. *Polyhedron* **13**(9), 1427 (1994).

170. Dvorkin, A.A.; Kokozay, V.N.; Petrusenko, S.R.; Simonov, Yu.A. *Ukr. Khim. Zh.* **57**(1), 5 (1991).

171. USSR Patent 1752735; *Bull. Izobr.* No. 29, 75 (1992).

172. Dvorkin, A.A.; Kokozay, V.N.; Petrusenko S.R.; Sienkiewicz, A.V. *Dokl. Akad. Nauk Ukr. SSR, Ser. B* No. 10, 30 (1989).

173. Sobolev, A.N.; Shvelashvili, A.E.; Kokozay, V.N.; Petrusenko, S.R.; Kapshuk, A.A. *Soobshch. Akad. Nauk GSSR.* **136**(2), 325 (1989).

174. Kokozay, V.N.; Petrusenko, S.R.; Sienkiewicz, A.V.; Simonov, Yu.A. *Ukr. Khim. Zh.* **59**(3), 236 (1993).

175. Kokozay, V.N.; Simonov, Yu.A.; Petrusenko, S.R.; Dvorkin, A.A. *Zh. Neorg. Khim.* **35**(11), 2839 (1990).

176. Simonov, Yu.A.; Skopenko, V.V.; Kokozay, V.N.; Petrusenko, S.R.; Dvorkin, A.A. *Zh. Neorg. Khim.* **36**(3), 625 (1991).

177. Vassilyeva, O.Yu.; Kokozay, V.N.; Skopenko, V.V. *Dokl. Akad. Nauk Ukr. SSR, Ser. B* No.8, 35 (1990).

178. Vassilyeva, O.Yu.; Kokozay, V.N.; Simonov, Yu.A. *Dokl. Akad. Nauk Ukr. SSR, Ser. B* No.10, 138 (1991).

179. USSR Patent 1641823; *Bull. Izobr.* No. 14, 90 (1991).

180. Kokozay, V.N.; Polyakov, V.R.; Simonov, Yu.A.; Sienkiewicz, A.V. *Zh. Neorg. Khim.* **37**(8), 1810 (1992).

181. Kokozay, V.N.; Sienkiewicz, A.V. *Polyhedron* **12**(19), 2421 (1993).

182. Kokozay, V.N.; Sienkiewicz, A.V. *J. Coord. Chem.* **30**(3–4), 245 (1993).

183. Kokozay, V.N.; Sienkiewicz, A.V. *J. Coord. Chem.* **31**(1), 1 (1994).

184. Vassilyeva, O.Yu.; Kokozay, V.N.; Skopenko, V.V. *Ukr. Khim. Zh.* **60**(3–4), 227 (1994).

185. Skopenko, V.V.; Kokozay, V.N.; Polyakov, V.R.; Sienkiewicz, A.V. *Polyhedron* **13**(1), 15 (1994).

186. Sienkiewicz, A.V.; Kokozay, V.N. *Z. Naturforsch., Teil B* **49**, 615 (1994).

187. Sienkiewicz, A.V.; Kokozay, V.N. *Polyhedron* **13**(9), 1431 (1994).

188. Kokozay, V.N.; Sienkiewicz, A.V. *Polyhedron* **13**(9), 1439 (1994).

189. Ivanova, E.I.; Kokozay, V.N. *Dokl. Akad. Nauk Ukr. SSR, Ser. B* No. 9, 135 (1994).

190. Kokozay, V.N.; Dvorkin, A.A.; Vassilyeva, O.Yu.; Sienkiewicz, A.V.; Rebrova O.N. *Zh. Neorg. Khim.* **36**(6), 1446 (1991).

191. Kokozay, V.N.; Petrusenko, S.R.; Simonov, Yu.A. *Zh. Neorg. Khim.* **39**(1), 81 (1994).

192. Prokofiev, A.I.; Prokofieva, T.I.; Bubnov, N.N.; Solodovnikov, S.P.; Belostotskaya, I.S.; Ershov, V.V.; Kabachnik, M.I. *Dokl. Akad. Nauk SSSR* **245**, 1393 (1979).

193. Volieva, V.B.; Prokofiev, A.I.; Prokofieva, T.I.; Ivanova, E.B.; Zhorin, V.A.; Ershov, V.V.; Enikolopyan, N.S. *Izv. Akad. Nauk SSSR, Ser. Khim.* 2800 (1986).

194. Annan, T.A.; McConvili, D.H.; McGarvey, B.R.; Ozarovski, A.; Tuck, D.G. *Inorg. Chem.* **28**, 1644 (1989).

195. Annan, T.A.; Tuck, D.G. *Can. J. Chem.* **67**, 1807 (1989).

196. Ozarovski, A.; McGarvey, B.R.; Peppe, C.; Tuck, D.G. *J. Am. Chem. Soc.* **113**, 3288 (1991).

197. Ceneschi, A.; Dei, A.; Gatteshi, D. *J. Chem. Soc., Chem. Commun.* 630 (1992).

198. Adams, D.M.; Reingold, A.L.; Dei, A.; Hendrickson, D.N. *Angew. Chem.* **105**, 434 (1993).

199. Pierpont, C.G.; Buchnan, R.M. *Coord. Chem. Rev.* **38**, 45 (1981).

200. Kaim, W. *Coord. Chem. Rev.* **76**, 187 (1987).

201. Pierpont, C.G.; Larsen, S.K.; Bose, S.R. *Pure Appl. Chem.* **60**, 1331 (1988).

202. Muraev, V.A.; Abakumov, G.A.; Razuvaev, G.A. *Dokl. Akad. Nauk SSSR* **217**, 1083 (1974).

203. Muraev, V.A.; Abakumov, G.A.; Razuvaev, G.A. *Dokl. Akad. Nauk SSSR* **236**, 3 (1977).

204. Prokofiev, A.I.; Malisheva, N.A.; Bubnov, N.N.; Solodovnikov, S.P.; Kabachnik, M.I. *Dokl. Akad. Nauk SSSR* **252**, 370 (1980).

205. Sofen, R.S.; Ware, D.C.; Copper, S.R.; Raymond, K.N. *Inorg. Chem.* **18**, 234 (1979).

206. Abakumov, G.A.; Muraev, V.A.; Razuvaev, G.A. *Dokl. Akad. Nauk SSSR* **215**, 1113 (1974).

207. Muraev, V.A.; Cherkasov, V.K.; Abakumov, G.A.; Razuvaev, G.A. *Dokl. Akad. Nauk SSSR* **236**(3), 620 (1977).

208. Tian, Z.; Tuck, D.G. *J. Chem. Soc., Dalton Trans.* 1381 (1993).

209. Annan, T.A.; Ozarowski, A.; Tian, Z.; Tuck, D.G. *J. Chem. Soc., Dalton Trans.* 2931 (1992).

210. a) Thomas, A.L. *Phthalocyanines. Research and Application.* Boca Raton: CRC Press (1990); b) Moser, F.H., Thomas, A.L. *Phthalocyanines. Properties*, Vol. 1. Boca Raton: CRC Press (1983).

211. *Phthalocyanines. Properties and Application* (Eds: Leznoff, C.C., Lever, A.B.P.). Weinheim: VCH (1993).

212. Kasuda, K.; Tsutsui, M. *Coord. Chem. Rev.* **32**, 67 (1980).

213. Kubiak, R.; Janszak, J. *J. Alloys Compounds.* **200**, L7–L8 (1993).

214. Abakumov, G.A.; Cherkasov, V.K.; Lobanov, A.V.; Razuvaev, G.A. *Izv. Akad. Nauk SSSR, Ser. Khim.* **7**, 1610 (1984).

215. Canadian Patent 2031707 (1989).

216. Dent, C.E.; Linstead, R.P. *J. Chem. Soc.* 1027 (1934).

217. Petit, M.A.; Plichon, V.; Belkacemi, H. *New J. Chem.* **13**, 459 (1989).

218. Tomoda, H.; Saito, S.; Ogawa, S.; Shiraishi, S. *Chem. Lett.* 1277 (1980).

219. Byrne, G.T.; Linstead, R.P.; Lowe, A.R. *J. Chem. Soc.* 1017 (1934).

Chapter 4

Mechanosynthesis of Coordination Compounds

4.1 INTRODUCTION

Among various methods of activation of chemical reactions, mechanical activation has its own characteristics and it is used, in general, for increasing the chemical transformation of inorganic and high-molecular compounds [1,2]. An increase in the velocity of metal dissolution, provoked by mechanical treatment, was first observed as long ago as the end of the 1920s [3]. In this respect, it is expected that an immediate mechanical activation of the "compact metal–ligand" system could contribute to the complex-formation process. The necessity of taking into account the mechanical energy involved in the synthesis of coordination compounds was known long ago and has been regularly studied [4,5].

The effects accompanying frictional interaction of solid substances are, undoubtedly, a primary cause of activation of the system "compact metal–ligand" under friction conditions. Taking into account Tissen's [6] arguments, according to which a "triboplasma" ("magmaplasma") appears in the contact zone, post-plasma effects [7,8] and activation of either the metal or the ligand, this seems very probable. That is, mechanical treatment of compact metal surfaces in the presence of complex-forming compounds could lead to the formation of various complexes of different compositions and structures.

There are not many literature data on the interaction between compact metals and aggressive nonaqueous media under friction (tribocorrosion in nonaqueous media), except a study on the acceleration evident in the corrosive destruction of metals, and characterizations (in some cases) of products formed as a result of tribocorrosion. A detailed study of the kinetics and mechanisms of triboprocesses is especially required.

Approximately 20 years ago [9], it was established that under friction of compact transition metals in nonaqueous solutions of proton-containing ligands, the main tribochemical reaction leads to the formation of a complex appearing at the friction surface and accumulating in the solution (if it is soluble in the solvent used) [10]. Complexes which are not formed using traditional (conventional) synthetic methods were obtained.

Further studies [12,13] have shown that preparative applications of this technique are restricted. However, the results are extremely important in the analysis [14–18] of the mechanisms of lubricant action in the processes of friction and wear of solid bodies, in particular of friction in conditions of self-organization of frictional systems (for example, in the regime of selective transfer [19,20]). In this respect, the present studies on complex formation by compact metals under friction are important first of all to develop the theory and practice of modern tribology.

4.2 COMPLEX FORMATION IN THE METAL–CARBON TETRACHLORIDE–DIMETHYL SULFOXIDE SYSTEM

It is known [21,22] that iron and copper are easily dissolved in mixtures of DMSO and CCl_4 forming complexes with solvated DMSO molecules in the coordination sphere. In the case of iron, depending on the reaction conditions, a series of complexes is formed in which the central atom has different oxidation states. In the case of copper, $[CuCl_2(DMSO)_2]$ was the only complex to be isolated.

Taking into account that a basic act of heterogeneous chemical reaction is localized in the surface of phase separation and that friction leads to the activation of this surface, the influence of the frictional conditions on iron and copper dissolution in the DMSO–CCl_4 system was studied [23].

This interaction was carried out in a special reactor (Fig. 4.1) [23] at 25°C, with a loading of 0.931 MPa for 2.5 h; the velocity of relative sliding of the samples was $0.125\,m\,s^{-1}$. The initial samples were prepared from copper M1, bronze AZMC or steel 45, and the counterbodies from bronze AZMC or steel 45. Pure CCl_4 and DMSO, and a CCl_4/DMSO mixture in a 1:1 volume ratio, were used as lubricant media. Wear, including the usual metal dissolu-

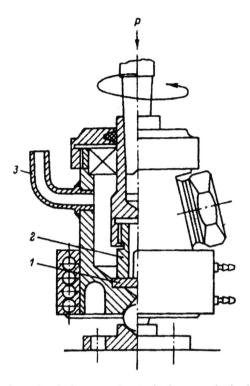

Fig. 4.1 Mechanochemical reactor: 1, sample; 2, counterbody; 3, refrigerator.

tion, was measured by a gravimetric method. The results are presented in Table 4.1.

As a result of the friction between bronze and bronze, or copper and steel, in the mixture of CCl_4 and DMSO, a slightly green sediment was isolated which corresponds to the $[CuCl_2(DMSO)_2]$ (m.p.152°C) reported earlier [22]. According to Table 4.1 the velocity of its formation is at least five times greater than in static conditions. Attempts to isolate an iron complex from solution in friction conditions using copper (or its alloy) and steel were unsuccessful. A copper complex is formed mainly as a result of a reaction, although the formation of iron complex also tales place under friction of steel against steel.

It is seen from Table 4.1 that complex formation by metals in the above system is considerably accelerated in friction conditions due to the formation of surface structures providing easier movement and separation of metal

TABLE 4.1
Influence of friction on the interaction between metals or their alloys with CCl_4 and DMSO

Friction pair	Medium	Friction coefficient	Wear of friction pair, g	Solubility, g/100 cm^3
Cu–bronze	CCl_4	0.31	0.0486	0
Steel–steel	CCl_4	0.39	0.0032	0
Cu–steel	CCl_4	0.29	0.0113	0
Bronze–bronze	DMSO	0.18	0.2637	0
Steel–steel	DMSO	0.42	0.0158	0
Cu–steel	DMSO	0.10	0.0051	0
Bronze–bronze	CCl_4 + DMSO	0.06	0.9077	0.0143
Steel–steel	CCl_4 + DMSO	0.25	0.0178	0.0012
Cu–steel	CCl_4 + DMSO	0.19	1.2441	0.0396

atoms. In the case of iron, the acceleration of complex formation is not as significant as in the case of copper. It could be explained by structural peculiarities of the coordination knot formed on the metal surface; the reaction is accelerated, in general, at the expense of an increase in the surface as a result of steel dispersion in the process of its wear. Such an idea is confirmed by the results of measuring friction coefficients. In the pairs with copper or its alloys, the decrease in the friction coefficient, (undoubtedly) related to complex formation, is more considerable than in the "steel–steel" friction pair. That is, the tribochemical reaction itself favorably influences the anti-friction properties of friction pairs. Such an influence probably becomes decisive in any case where there are definite configuration conditions for reaction products on the friction surface.

Thus, friction accelerates the process of complex formation of iron and copper in CCl_4–DMSO mixtures. A copper complex is the principal reaction product; an iron complex is not formed or it is formed in small quantities. Besides the causes stated above, this could be related to additional activation of copper and its "deformation", as a softer metal, in friction conditions.

4.3 INTERACTION OF COPPER WITH SALICYLIC ALDEHYDE

In friction between copper and copper with salicylic aldehyde solutions in decane, toluene or ethanol, formation of copper(II) *bis*-salicylalaldehydate takes place (reaction (4.1)) [24]:

TABLE 4.2
Influence of solvent on interaction of copper with salycylic aldehyde

Solvent	Interaction type	UV absorption λ_{max}, nm	$C_c \times 10^4$, mol dm^{-3}	$W \times 10^4$, mol dm^{-3} cm^{-2}
Ethanol	T	385	2.30	20.0
Ethanol	C	385	2.00	6.0
Toluene	T	390	0.37	11.3
Toluene	C	390	0.20	0.6
Decane	T	390	0.10	4.7
Decane	C	390	0.03	0.09

C_c, complex concentration in reaction mixture; W, reaction yield per surface unit.

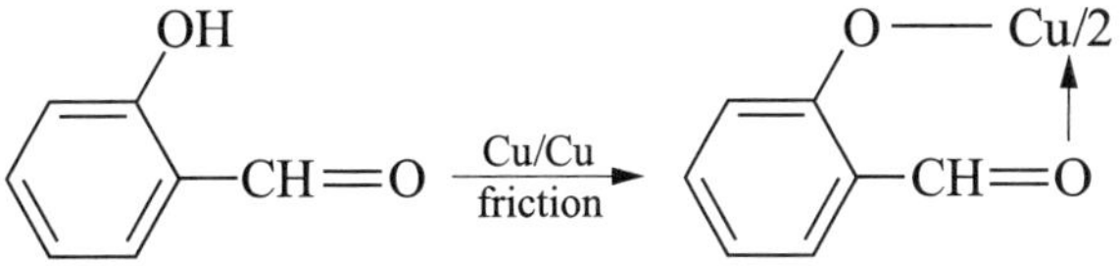

$$(4.1)$$

The same product is formed as a result of the interaction between compact copper and salicylic aldehyde solutions. Comparative results on the chemical (C) and tribochemical (T) processes are presented in Table 4.2.

From these data it can be seen that the reaction rate and yield of the complex are always higher in tribochemical interactions than by the chemical route. An increase in the polarity of the solvent and its solvating capacity leads to the intensification of both the chemical and tribochemical processes. The "tribochemical yield/chemical yield" ratio is increased in the order ethanol < toluene < decane. Supposing that the complex formed is fixed to the copper surface, forming a border layer, its transfer into the solution is linked with the destruction of this layer, and it becomes clear why the difference in complex accumulation under friction conditions and in chemical interaction is smaller in ethanol (which is the most effective among the solvents studied) than in toluene and decane. In fact, since salicylic aldehyde in decane practically does not react with copper in normal conditions, a maximum difference between W_T and W_C in this case testifies that the complex is formed exclusively in the friction contact zone.

On the basis of the above results, it is possible to conclude that the ratio of the amounts of complexes formed in both types of interaction (tribochemical and chemical) is the higher, the lower the polarity and solvating capacity of solvent.

4.4 INTERACTION OF COPPER WITH o-AZOMETHINIC LIGANDS

The first experimental proof of complex formation was obtained in studies on the interaction between some o-azomethines and copper by bronze in friction conditions [9,11,13,25–27]. It was established that, after 30 min of friction in 1% ethanol solutions of the ligands, the corresponding chelates (reactions (4.2)–(4.6)) are formed in 30% yield; they have melting points, UV and IR spectral data corresponding to those reported in the literature [28].

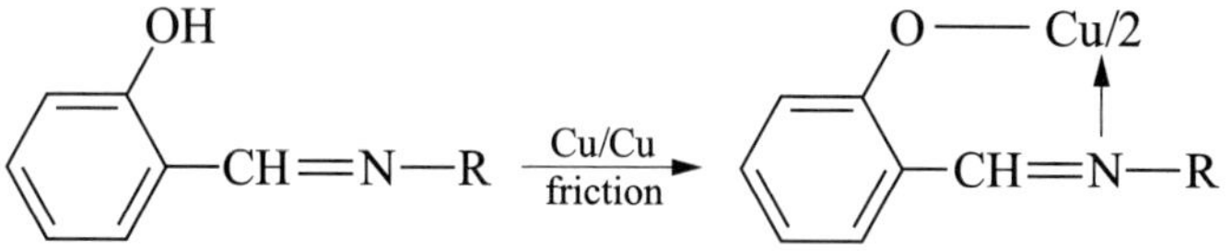

$$\text{(4.2)}$$

R = Et, Bu, C_6H_{13}, 4-$N(CH_3)_2$-C_6H_4, 4-OCH_3-C_6H_4, 4-CH_3-C_6H_4, 4-C_6H_5, 4-Cl-C_6H_4, 4-Br-C_6H_4, 4-OH-C_6H_4, 4-NO_2-C_6H_4, 2-OCH_3-C_6H_4, 2-CH_3-C_6H_4, 3-CH_3-C_6H_4, 3-NO_2-C_6H_4

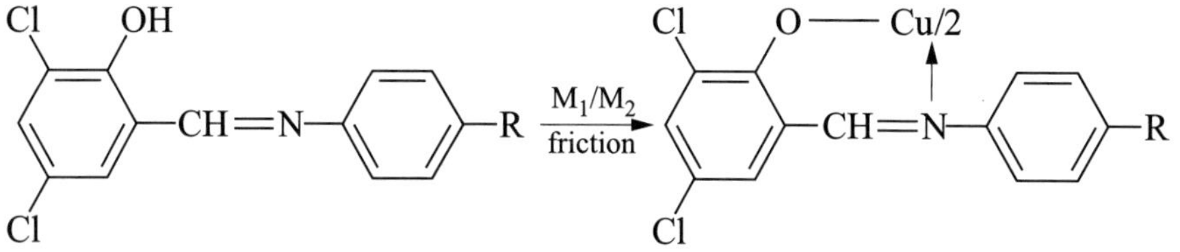

$$\text{(4.3)}$$

R = OCH_3, CH_3, Br, NO_2, H

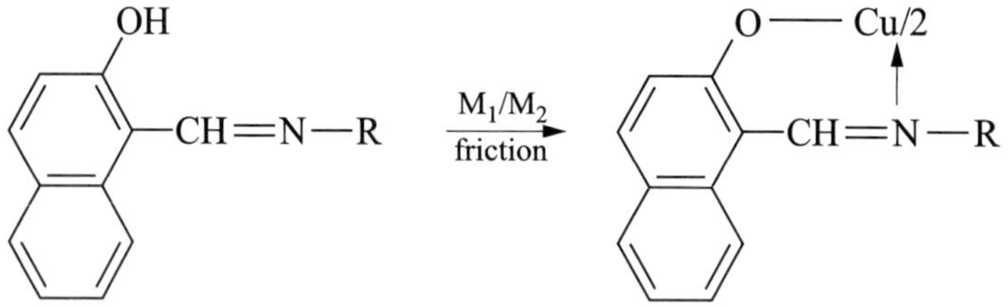

$$\text{(4.4)}$$

R = Ph, 4-CH_3-C_6H_4, 4-OCH_3-C_6H_4, 4-I-C_6H_4, 4-OH-C_6H_4, $C_{10}H_7$, Py

$$(4.5)$$

$$R_1 = H, R_2 = H; \; R_1 = H, R_2 = N(CH_3)_2; \; R_1 = N(CH_3)_2, R_2 = NO_2$$

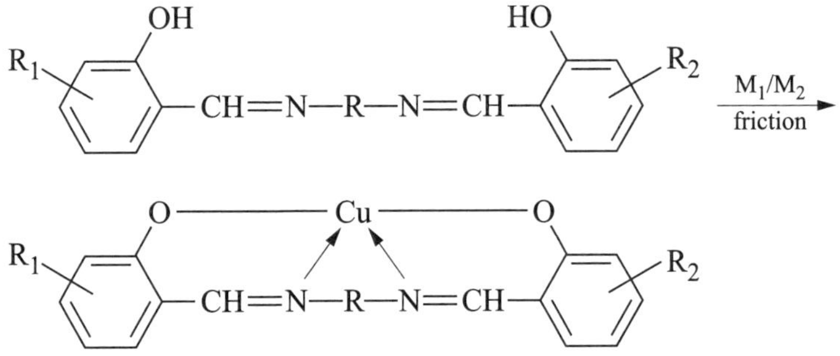

$$(4.6)$$

$$R = -CH=CH-, \; R_1 = R_2 = H; \; R = 1, 2\text{-}C_6H_4, \; R_1 = R_2 = H$$

It was shown that the interaction between copper powder and a boiling solution of the ligand (1 h) without friction leads to the appearance of a typical colored copper complex. However, the product formed could not be isolated because of its low concentration. Thus, the presence of the hydroxy-group in the *o*-hydroxyazomethine ligand leads to a considerable (dozen- and hundred-fold) acceleration of the complex-formation processes in comparison with the direct synthesis in solution.

Under friction between copper and copper, copper and steel, bronze and bronze, bronze and steel, and nickel and steel (solvents: ethanol, DMF, vaseline oil or polyorganosiloxanes), copper and nickel complexes with a series of *o*-oxyazomethinic ligands have been obtained; their composition and properties are the same as those of the corresponding complexes obtained by conventional synthetic methods [29]. At the same time, a tribo-chemical reaction in CCl_4 could lead, in some cases, to the formation of chloro-containing products. Substitution of the hydrogen atom of the hydroxy group in compounds **I–IV** by another functional group leads to the cessation of the interaction between the metal and the ligand both in the usual conditions of direct synthesis and in friction conditions. The spectra of solutions of these compounds in the visible range do not differ from the spectra of initial mixtures. It is interesting that, as in the previous systems, with friction of copper and its alloys against steel it is always only copper complexes that are observed to be formed.

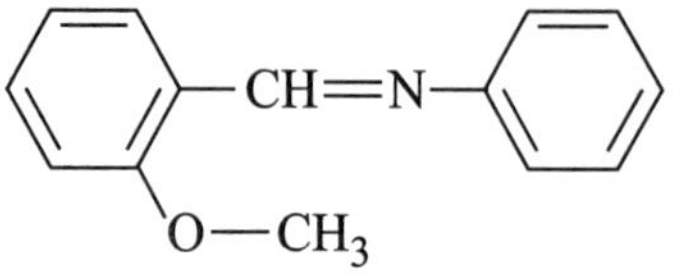

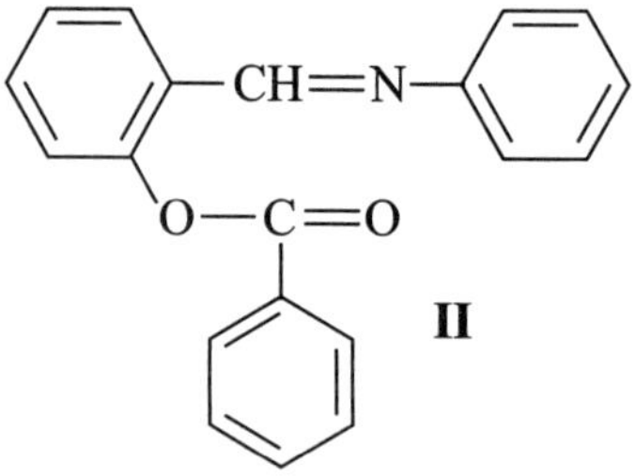

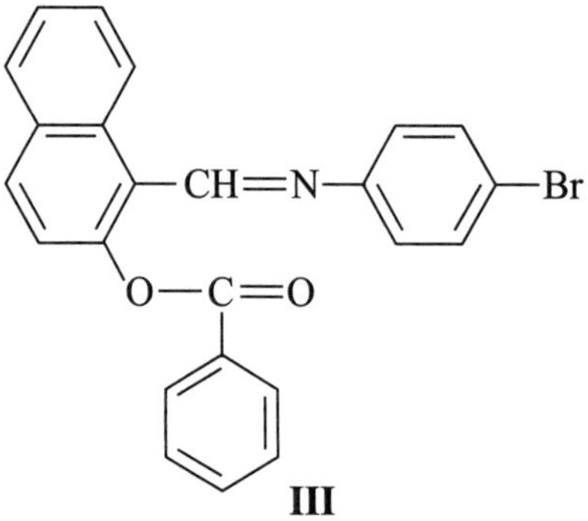

Thus, it has been established that tribochemical transformations with *o*-oxyazomethines in friction pairs with copper lead to the formation of copper chelates as the main reaction products which are accumulated in solution, and they could be isolated individually. Formation of a stable coordination knot CuO_2N_2 with a five- or six-membered quasi-aromatic ring is possible in the case of a hydroxy group in the *o*-position in relation to the azomethinic bond.

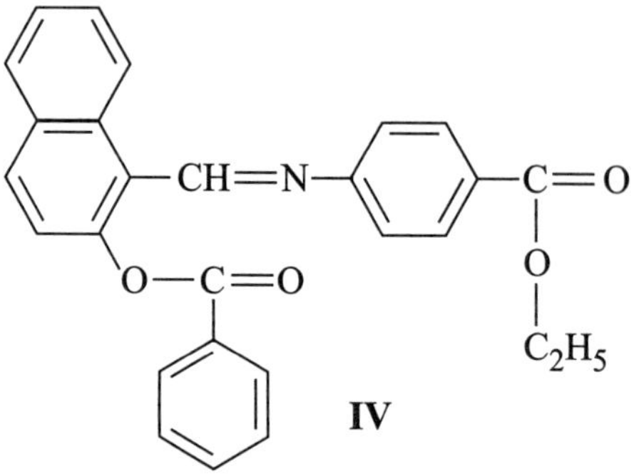

4.5 INTERACTION OF COPPER WITH SCHIFF BASES OF 1-PHENYL-3-METHYLPYRAZOL-5-ONE AND ITS SULFUR- AND SELENIUM-CONTAINING ANALOGUES

Interaction between copper, in the pair copper–steel, and the above ligands in friction conditions leads to various products, depending on the solvent and ligand used [29]. Thus, chelates which are analogous to those of the azomethines are formed in ethanol by reaction (4.7).

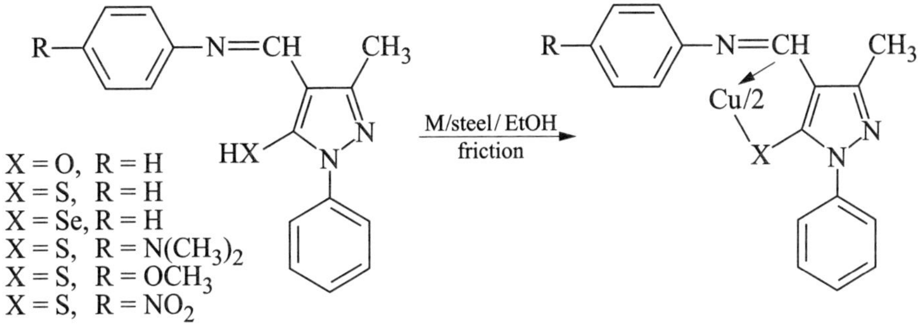

$$X = O, R = H$$
$$X = S, R = H$$
$$X = Se, R = H$$
$$X = S, R = N(CH_3)_2$$
$$X = S, R = OCH_3$$
$$X = S, R = NO_2$$

$$\xrightarrow{\text{M/steel/EtOH} \atop \text{friction}}$$

(4.7)

At the same time, the products obtained in CCl_4 and acetone differ from each other and from those obtained in ethanol. The greatest differences are observed in case of S- and Se-containing ligands: their reaction products have different IR spectra and color. Thus, green crystalline powders were isolated from acetone, and brown powders from CCl_4. The IR spectra of compounds isolated from acetone showed, additionally to the absorption band ν(C–N) which moved by 5–15 cm^{-1} to lower frequency, an absorption band ν(CO) at 1680–1690 cm^{-1}, which testifies to the presence of solvent molecules in the coordination sphere of the complex. So, if the reaction in ethanol could be described by equation (4.8), the corresponding reaction in acetone leads to an additional product, the complex $CuL_2 \cdot nC_3H_6O$.

$$Cu + 2HL + \tfrac{1}{2}O_2 = CuL_2 + H_2O \qquad (4.8)$$

The reaction in CCl_4 leads to amorphous chloro-containing compounds of variable composition. According to IR data, they contain azomethinic and pyrazolone fragments. Probably in this case the compounds $Cu_nL_mCl_k$ are formed, which, by analogy with the CCl_4–DMSO system, have a composition depending on the conditions of friction.

4.6 INTERACTION OF COPPER WITH o-HYDROXYAZO COMPOUNDS

The presence of the $-N{=}N-$ group in azo compounds contributes to the formation of both five- and six-membered metal-containing rings in their interaction with transition metals (reaction (4.9)) [30]:

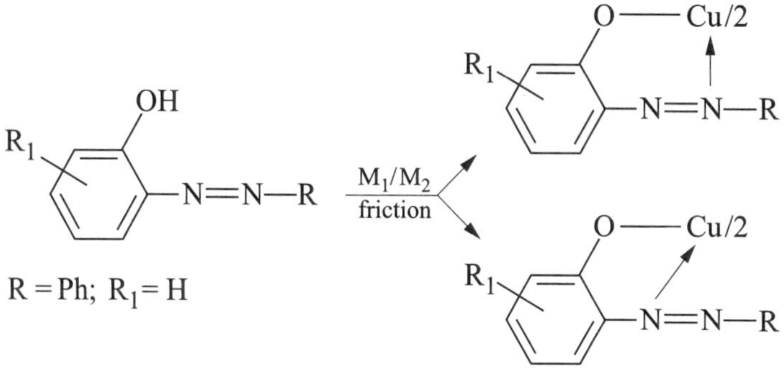

$$(4.9)$$

$R = Ph; \ R_1 = H$

Experimental results for friction of copper by steel in ethanol solutions of hydroxyazo compounds showed [10,11] that these ligands, as well as the azomethines, form mostly 1:2 complexes with the coordination knot MN_2O_2. Only in the case of 1-pyridyl-2-azonaphthol, is the 1:1 complex formed; this is related to the presence of a third coordination center in the pyridine fragment of the ligand. Evidently the fourth place in the coordination sphere of the complex is occupied by the solvent molecule.

In the case of the hydroxyazomethines, also, blocking of the hydroxy group leads to the absence of interaction in the systems, although in the case of the S-containing analogues of hydroxyazo compounds **V** and **VI** a mixture of unidentified copper complexes is formed.

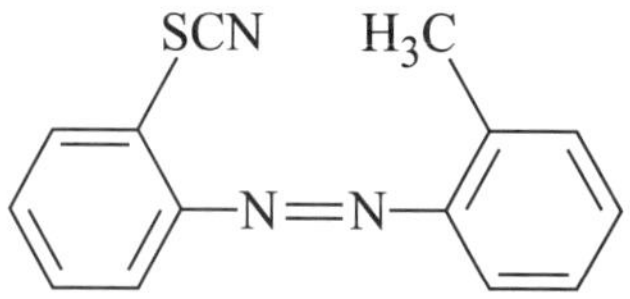

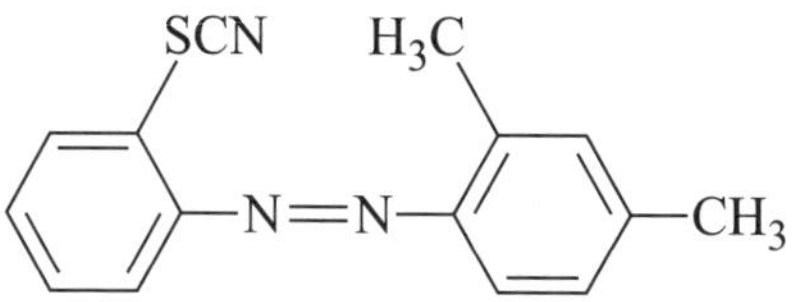

4.7 INTERACTION OF COPPER AND NICKEL WITH 8-HYDROXYQUINOLINE AND ITS DERIVATIVES

Tribochemical reactions with participation of 8-hydroxyquinoline and its derivatives has a possible practical aspect of the possibility of their use as a component of lubricant materials [11,27,31]. According to the reported data [11,27], the interaction of 8-hydroxyquinoline with copper and nickel under friction between copper or nickel with steel leads to the formation of the corresponding complexes (reaction (4.10)). Iron complexes are not formed in these conditions.

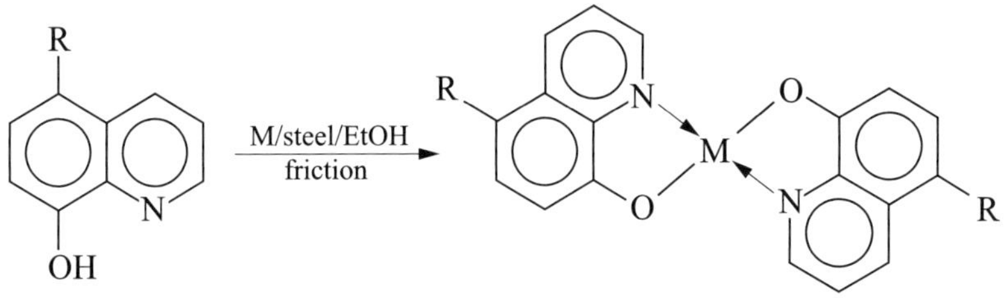

R = H, Cl, Br

$$(4.10)$$

The copper and nickel complexes of 8-hydroxyquinoline and its derivatives, in contrast o-oxyazomethines and o-hydroxyazo compounds, are also formed when a hydrogen atom in the OH group of the ligand is substituted by a propyl radical. In spite of the almost double activation time of nickel in comparison with copper, the yield of nickel complexes (10–20%) is considerably lower than the yield of copper complexes (30–40%) [27,29].

4.8 INTERACTION OF COPPER AND IRON WITH ACETYLACETONE AND 1-(TENOIL-2)-3,3,3-TRIFLUOROACETONE

Acetylacetone and its derivatives are interesting ligands from viewpoint of modeling the products of the tribochemical oxidation of hydrocarbons, whose lubricant properties are improved if oxo- or hydroxy groups are present in their molecules [32].

A study of solutions of acetylacetone and 1-(tenoil-2)-3,3,3-trifluoroacetone in decane or DMF before and after friction of copper against copper, copper against steel, and steel against steel made it possible to isolate the corresponding chelates (reaction (4.11)). Data confirming formation of copper chelates were obtained after isolation of the complexes from solution and further physicochemical studies [29,33].

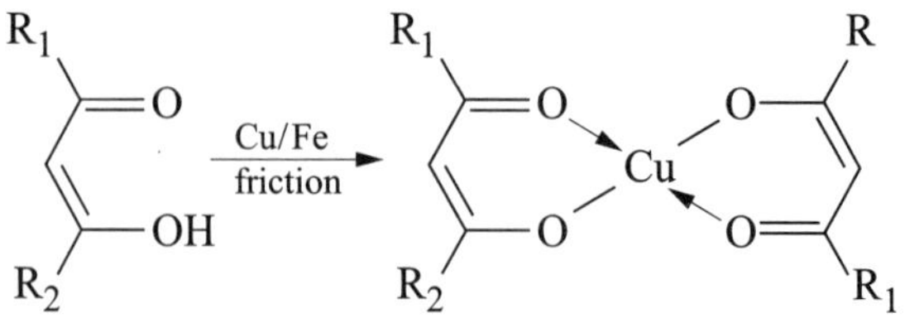

$$R_1 = R_2 = CH_3; \; R_1 = CF_3, \; R_2 = C_4H_3S$$

(4.11)

In the case of acetylacetone under friction of steel against steel, a mixture of iron(II) and iron(III) complexes is formed [29]. The results obtained by friction of copper against steel in acetylacetone solutions in decane [29] are especially interesting. In this case, the formation of copper and iron complexes was observed; their ratio changed according to the reaction conditions. Thus, the friction in the copper–steel pair during 0.5 h changed the initial colorless acetylacetone solution to red (iron complex). In 1.5 h a violet sediment of iron complex was isolated; the solution became blue and contained only a blue complex.

On the basis of these results, it is possible to suppose that analogous tribochemical transformations could take place in the case of other β-diketones. The presence of oxygen (air) could also influence the lubricant properties of hydrocarbons and other organic substances in frictional contact [34].

In spite of the presence of other active centers (S and F) in 1-(tenoil-2)-3,3,3-trifluoroacetone, the observed tribochemical reactions do not lead to the formation of copper or iron sulfides and fluorides, which traditionally have an especial role in the formation of a border layer of friction.

4.9 INTERACTION OF COPPER WITH *N*-ACYLSALICYLAMIDE SOLUTIONS

The presence of four donor centers in *N*-salicylamides makes it feasible to propose [35] the possibility of formation of four types of chelate complexes (scheme (4.12)). An interaction of these ligands in conditions of friction of copper against copper was studied in DMF solutions [36]. In all cases, the acceleration of the reactions was considerably higher than in the case of hydroxyazomethines or hydroxyazo compounds. This is probably related to the lower activity of *N*-salicylamides in direct synthesis conditions and their higher susceptibility to activation. The composition and structure of the products correspond to the type B. The observed yields of the variants A, C, D, were higher than those theoretically calculated.

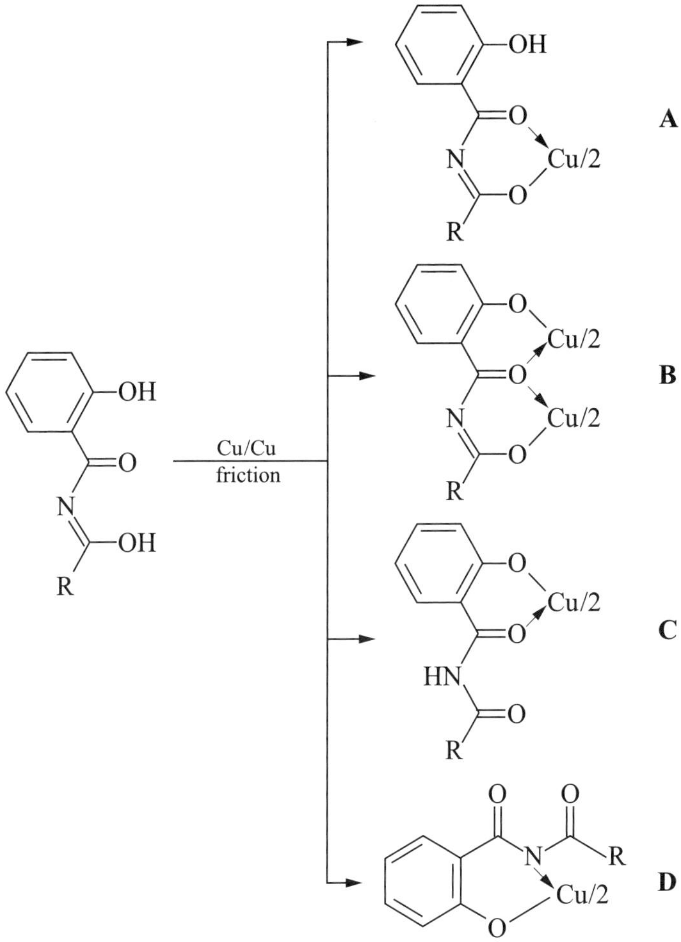

$$(4.12)$$

R = C_4H_9, C_5H_{11}, Ph, $CH_2C_6H_5$, $CH{=}CH{-}Ph$, $4{-}OCH_3{-}C_6H_4$, $4{-}CH_3{-}C_6H_4$, $4{-}HO{-}C_6H_4$, $4{-}Cl{-}C_6H_4$, $4{-}Br{-}C_6H_4$, $3{-}NO_2{-}C_6H_4$

It is noteworthy that an increase of reaction time to 6h leads to the formation of two other probably mononuclear complexes **A** and **C** with $\lambda_{max} = 360$ and 420 nm respectively. This testifies to the possibility of participation of the primarily formed binuclear complex in further reactions.

4.10 INTERACTION OF COPPER WITH DISALICYLAMIDE

Under friction of copper against steel in ethanol solutions of disalicylamide, small crystals of green precipitate, practically insoluble in alcohols, ether and CCl_4, and slightly soluble in DMF, are formed [29]. The results of a physicochemical study showed that the trinuclear copper complex with disalicylamide reported earlier [37] is formed (reaction (4.13)).

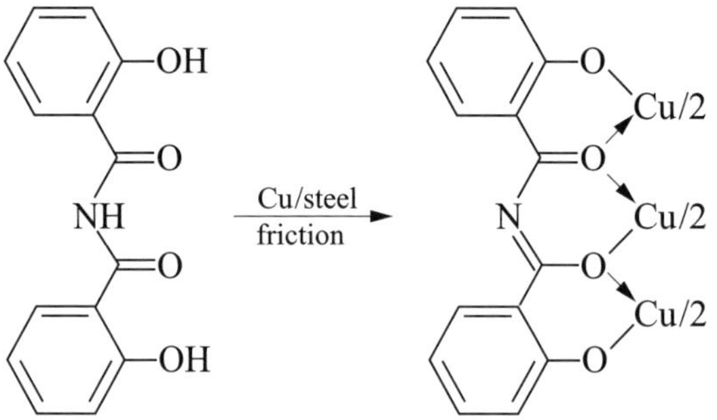

$$(4.13)$$

Thus, under friction conditions, in addition to mononuclear complexes, polynuclear ones could be formed; their composition and structure are determined by the structure of the initial ligands. The following circumstance plays an important part: the complex obtained contains a linear chain in which the distance between copper atoms is 2.58 Å, close to the shortest distance between atoms in the crystal structure of elemental copper [38]. This fact makes possible the participation of not only individual atoms, but also surface fragments, in complex formation, since successive entry of copper atoms into the complex is unlikely due to its low solubility.

4.11 FORMATION OF COORDINATION METAL–POLYMERS

An idea [39] about the possibility of formation of organic polymers under friction conditions was used to obtain metal–polymers [29]. It was shown that the interaction of ethanol solutions of di-*o*-hydroxyphenyl-1,2,4-oxadiazole and some 1,2,4-triazoles under conditions of friction of copper against steel leads to the formation of products with a high melting point ($> 280°C$) which are insoluble in organic solvents (reaction (4.14)). Their physicochemical characteristics correspond to those of the polymer complexes obtained by conventional chemical routes [40].

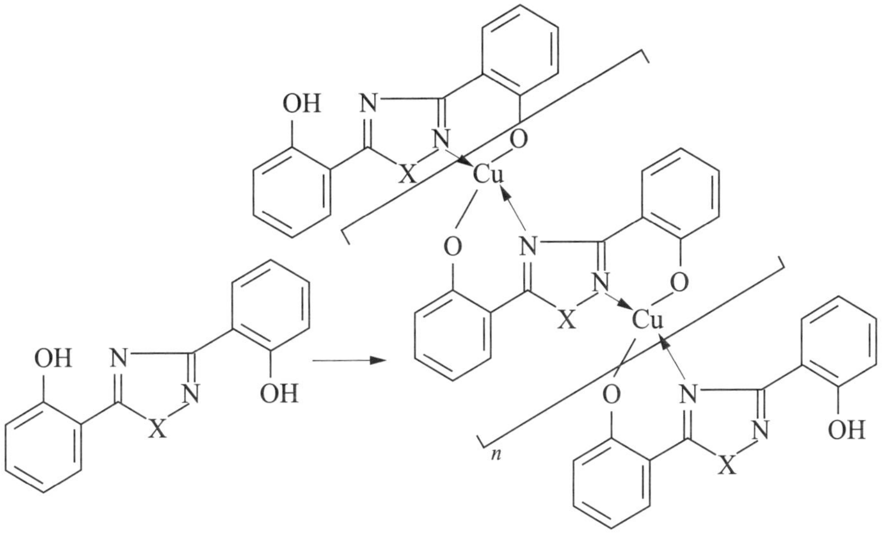

(4.14)

This opens the possibility of explaining analytical results reported in [41], where the polymer obtained contained 72.7% C, 6.6% H, and 4.1% Fe. The presence of the second absorption band $\nu(CO)$, moved to the high-wavelength area, could be explained by coordination of the carboxyl group to the iron atom.

As shown from the results above, both mono- and polynuclear complexes can be the products of tribochemical reactions; their composition and structure depend on the structure of the initial ligands and the nature of the solvent. The tribochemical process is most effective when the ligand has an active hydrogen atom in an OH, SH, SeH or NH group.

Experimental procedures

Cu, Ni + azomethinic ligands using ultrasound treatment

Metal powder was put into the flask with ligand solution (10%) in EtOH or dioxane. The ultrasonic treatment was carried out using UZDN-1 equipment (35 kHz) for 1–3 h. X-ray powder diffraction was used for identification of the products **VII** [12].

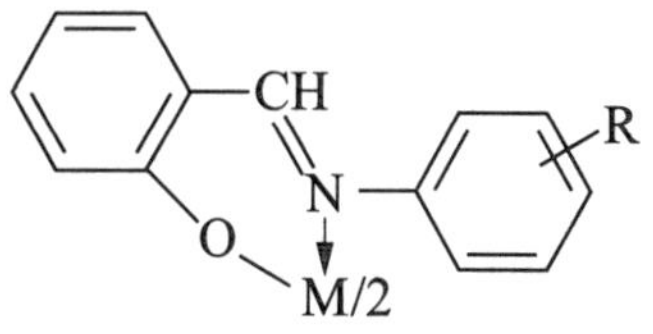

R = CH$_3$, H, Cl, NO$_2$
M = Cu, Ni

VII

Reactions of metal powders and azomethinic ligands also take place without ultrasonic treatment (in very small yields); however, it is impossible to identify the products using X-ray powder diffraction.

According to studies reported in [12], a multimolecular layer of the product is formed in the metal surface. Since for its formation the presence of metal atoms or ions at the border between liquid and solid phases is needed, diffusion of metal atoms through the compound layer is a necessary condition for such a layer to form. Cavitation processes on the surface contributes to it. Since an energy barrier has to be overcome on the reaction route, ultrasonic cavitation action has the same importance as triboplasma formed by friction of metals [42].

Synthesis of copper(II) salicylalanilinate

Two methods were compared.

In the first, an ethanol solution (1%) of salicylalaniline (10 cm^3) was treated in the reactor (Fig. 4.1) for 20 min at a velocity of relative sliding of 0.4 m s^{-1} and a loading 50 kg cm^{-2} using a copper (sample)–bronze (counterbody) friction pair. The brown solution (without solid) was filtered; the solid metal residues were washed with hot ethanol. Mother solutions were united and evaporated. The yield of brown product was 60% [20].

In the second method, a 0.1% solution of salicylalaniline in polymethylsiloxane (5 cm^3) was treated for 5 h in a reactor at a velocity of relative sliding of 0.8 m s^{-1} and a loading 50 kg cm^{-2} using a bronze–steel friction pair. The subsequent isolation was similar to the procedure described above. Yield 18% [29].

Synthesis of copper(II) salicylal-p-chloroanilinate

A 1% solution (10 cm^3) of salicylal-*p*-chloroaniline in EtOH was treated in the reactor similarly to the above method, with a copper (sample)–bronze (counterbody) friction pair. Yield 20% [29].

Synthesis of copper(II) benzal-o-aminophenolate

A 1% ethanol solution (10 cm^3) of salicylalaniline was treated in the reactor (Fig. 4.1) for 2 h at a velocity of relative sliding of 0.8 m s^{-1} and a loading 50 kg cm^{-2} using a copper (sample)–steel (counterbody) friction pair. The black solid formed was filtered and washed with hot ethanol. Yield 20% [29].

Synthesis of copper(II) benzeneazo-β-naphtholate

A 0.1% solution of benzeneazo-β-naphthol in vaseline oil was treated for 20 h in a reactor at a velocity of relative sliding of 0.003 m s^{-1} and a loading 100 kg cm^{-2} using a bronze–steel friction pair. The solid formed was filtered, washed with EtOH and dried. Yield 15% [29].

Synthesis of nickel(II) salicylalanilinate

A 1% ethanol solution (10 cm^3) of salicylalaniline was treated in the reactor for 40 min at a velocity of relative sliding of 0.8 m s^{-1} and a loading 50 kg cm^{-2} using a nickel (sample)–steel (counterbody) friction pair. The green solution formed was filtered and evaporated. Yield 15% [29].

4.12 CONCLUSIONS

From the results, the overall character of the evolution of organic compounds from their tribochemical transformation processes may be proposed. Evidently, the triboactivation of organic molecules leads to primary radical reactions of polymerization, followed by tribocoordination reactions and formation of inorganic compounds of metals (the undesirable results of tribochemical processes). Under friction conditions, the tribocoordination is an intermediate stage during evolution of the tribosystem. Tribochemical reactions are important in industrial processes related with friction and lubricants, and require more profound study.

REFERENCES

1. Avaakumov, E.G. *Mechanical Methods of Activation of Chemical Processes.* Novosibirsk: Nauka (1986).
2. Baranboim, N.K. *Mechanochemistry of High-Molecular Compounds.* Moscow: Mir (1971).
3. Tammam, G.Z. *Electrochemisty* **35**, 21 (1929).
4. Waneting, P. *Kolloid. Zh.* **41**, 152 (1927).
5. Berlin, A.A. *Usp. Khim.* **27**, 94 (1958).
6. Tissen, P.A.; Meyer, K.; Heinike, G. *Abh. Deutsch. Acad. Wiss.—Kl. Chem., Geol., Biol.* (1), 15 (1961).
7. Tissen, K.P. *Z. Phys. Chem. (Leipzig)* **260**(3), 403 (1979).
8. Tissen, K.P.; Sieber, K.Z. *Z. Phys. Chem. (Leipzig)* **260**(3), 410 (1979).
9. Kuzharov, A.S.; Barchan, G.P.; Chuvaev, V.V. *Zh. Fiz. Khim.* **51**(11), 2949 (1977).
10. Kuzharov, A.S.; Bolotnikov, V.S. *Zh. Fiz. Khim.* **53**(10), 2639 (1979).
11. Kuzharov, A.S.; Garnovskii, A.D.; Kutkov, A.A. *Zh. Obshch. Khim.* **49**(4), 861 (1979).
12. Kuzharov, A.S.; Suchkov, V.V.; Vlasenko, L.A. *Zh. Fiz. Khim.* **53**(8), 2064 (1979).
13. Kuzharov, A.S.; Suchkov, V.V. *Zh. Fiz. Khim.* **54**(12), 3114 (1980).
14. Grechko, V.O. Ph.D. Thesis. Novocherkassk (1982).
15. Danyushina, G.A. Ph.D. Thesis. Rostov-on-Don (1992).
16. Onishuk, N.Yu. Ph.D. Thesis. Novocherkassk (1983).
17. Ryadchenko, V.G. Ph.D. Thesis. Novocherkassk (1988).
18. Fisenko, O.V. Ph.D.Thesis. Rostov-on-Don (1994).
19. Garkunov, D.N. *Tribotechnique.* Moscow: Mashinostroenie (1985).
20. *Discoveries in USSR* (brief descriptions of 1957–1967). Moscow: TSNIIPI (1972), p. 52.
21. Lavrentiev, I.P.; Korableva, L.G.; Lavrentieva, E.A. et al. *Izv. Akad. Nauk SSSR, Ser. Khim.* **7**, 1961 (1974).
22. Kurskov, S.N.; Ivleva, I.N.; Lavrentiev, I.P. et al. *Izv. Akad. Nauk SSSR, Ser. Khim.* **9**, 1442 (1976).
23. Kuzharov, A.S.; Zhuravleva, S.A.; Shakurova, I.K. *Zh. Fiz. Khim.* **55**(11), 2872 (1981).
24. Kuzharov, A.S.; Sapelkina, N.P. *Memories of the 1st Conference on Physical and Organic Chemistry of North Caucasus,* Rostov-on-Don (1989), p. 82.
25. Kuzharov, A.S. *Thesis of the Friction Conference,* Dnepropetrovsk (1981), p. 70.
26. Kuzharov, A.S.; Garnovskii, A.D. *Thesis of XIV Conference on Coordination Chemistry,* Ivanovo (1981), p. 412.
27. Kuzharov, A.S.; Suchkov, V.V.; Vlasenko, L.A. et al. *Zh. Fiz. Khim.* **55**(10), 2588 (1981).
28. Garnovskii, A.D. *Russ. J. Coord. Chem.* **19**(5), 368 (1993).
29. Kuzharov, A.S. *Dr. Hab. Thesis,* Rostov-on-Don (1991).
30. Shkolnikova, L.M.; Shugam, E.A. *Results Sci. Technol., Sect. Crystallochem.* **12**, 169 (1977).
31. Chichibabin, A.E. *Bases of Organic Chemistry.* Moscow: GCI (1957).
32. Akhmatov, A.S. *Molecular Physics of Friction.* Moscow: Fiz.-Khim. Liter. (1963).
33. Kuzharov, A.S.; Suchkov, V.V.; Komarchuk, L.A. *Zh. Fiz. Khim.* **57**(7), 1748 (1983).
34. Aksenov, A.F. *Friction and Wear of Metals in Hydrocarbon Liquids.* Moscow: Mashinostr. (1977).

35. Ryabukhin, Yu.I. Dr.Hab. Thesis, Rostov-on-Don(1991).

36. Kuzharov, A.S.; Ryabukhin, Yu.I. *Trenie i Iznos.* **12**(1), 99 (1991).

37. Uflyand, I.E.; Faleeva, L.N.; Vysotsky, B.D. et al. *Koord. Khim.* **8**(4), 494 (1982).

38. Gorelik, S.S.; Rastorguev, L.N.; Skakov, Yu.A. *X-Ray and Electron-Optic Analysis.* Moscow: Metallurgia (1970).

39. Zaslavskii, Yu.S.; Zaslavskii, V.N. *Action Mechanism of Antiwear Additives for Lubricants.* Moscow: Khimiya (1978).

40. Ryabukhin, Yu.I.; Shibaeva, N.V.; Kuzharov, A.S. et al. *Koord. Khim.* 13(7), 869 (1987).

41. Feih, R.S.; Kreus, K.L. *News in Lubricants.* Moscow: Chemistry (1967).

42. Heinicke, G.; Fleischer, G. *Die Technik.* **31**, 458 (1976).

Chapter 5

Some Peculiarities of the Crystal Structures of Complexes Obtained by Direct Synthesis

As mentioned previously, one of the peculiar properties of the compounds obtained by direct synthesis is their uncommon composition, which is directly connected with the original steric structure. These peculiarities of crystal structures may be illustrated by the complexes obtained directly from metal powders or metal oxides and ammonium salt solutions. More than 30 structures have been obtained. Their basic crystallographic characteristics are listed in Table 5.1 (fragments of some of these structures are shown in Figs 5.1–5.4, below). To analyze the data obtained, one ought to pay attention to the states of the coordination polyhedron of the central atom and of the ligands. The compounds may contain different polyhedra of the same metal. Thus, the complexes $[Cu(En)_2]I_2 \cdot 0.33DMF$ [1], $[Ni(En)_3]I_2 \cdot 2DMSO$ [26] and $[Cu(HDea)(H_2O)](NO_3) \cdot 2H_2O$ [19] have two, and $[Pb(En)_2](SCN)_2$ has [4] three, coordination polyhedra with distinct geometrical parameters (En, ethylenediamine; Dea, doubly deprotonized diethanolamine).

Most complexes have two types of different polyhedra: a distorted square pyramid with a lead atom at an apex, and a trigonal pyramid ($[Pb_2(NCS)(HTea)(SCN)$ [21]) or bisphenoid with the lone pair of electrons in the equatorial plane ($[Pb_2(HTea)](Ac)_2$ [21]) (HTea, doubly deprotonized triethanolamine. $[Pb_2(NO_3)_2(HTea)(DMSO)]$ [21] has pseudo-octahedral coordination, and $[Pb_2(HTea)]I_2$ [22] contains a lead atom in a pseudo-trigonal environment. Complexes $[Pb_2(Dea)](NO_3)_2 \cdot DMF$ [18] and $(H_3Dea)[Pb_2Br_6]$ [18] have both a distorted trigonal pyramid with a lead atom at an apex and a bisphenoid or pseudo-trigonal type of coordination with the lone pair of electrons at one apex of the triangle (for $[Pb_2(Dea)](NO_3)_2 \cdot DMF$). $[Pb_5Cl_6(Me_2Ea)_4]$ [16] (Ea, deprotonized ethanolamine) contains the lead atoms in pseudo-octahedral and distorted dodecahe-

TABLE 5.1

X-ray data of some complexes obtained by direct synthesis

Formula	Space group	Unit cell parameters						Ref.
		a, Å	b, Å	c, Å	α	β	γ	
[Cu(En)$_2$]I$_2$·0.33DMF	$Bm2_1b$	21.501	15.770	12.016	90	90	90	1
[Ni(En)$_3$]I$_2$·2DMSO	$P2_1$	15.283	15.184	11.930	90	90	71.5	26
[Pb(Ac)$_2$(En)]	$P2_1/c$	5.532	24.185	7.630	90	101.3	90	2
[Pb(En)$_2$]Br$_2$	$A2/a$	13.183	13.000	7.542	90	90	62.2	3
[Pb(En)$_2$](SCN)$_2$	$C2/c$	19.657	14.725	24.478	90	97.4	90	4
[PbI(En)]I·DMSO	$P\bar{1}$	10.225	10.132	6.900	90.8	88.3	106.4	5
[Zn(En)$_3$](SCN)$_2$·KSCN	$P\bar{3}c1$	11.665	11.665	16.80	90	90	120	6
[Cd(En)$_3$](SCN)$_2$·KSCN	$P\bar{3}c1$	11.868	11.868	16.917	90	90	120	7
[Ni(En)$_3$](SCN)$_2$·NH$_4$SCN	$P\bar{3}c1$	11.960	11.960	17.145	90	90	120	8
[Zn((SCN)$_2$Den)$_2$]	$Pbca$	22.931	13.016	10.545	90	90	90	9
[Ni(HTen)$_2$(SCN)$_4$]	$P2_1/n$	14.438	7.204	11.214	90	90	96.6	10
[Ni(SCN)$_2$(H$_2$O)$_2$·(Ten)$_2$]·4DMF	$P2_1/a$	12.783	11.263	12.910	90	90	71.5	11
[Cu(Ea)(HEa)$_2$]I	$p2_1/b$	11.963	10.178	11.470	90	90	114.7	12
[Cu(Ea)(SCN)(HEa)]	$P\bar{1}$	8.506	10.579	12.663	77.4	72.9	64.9	13
[Pb$_6$(Ea)$_6$I(HEa)$_2$]I$_5$	$P\bar{1}$	15.990	14.32	11.790	90	88.5	60	14
[Pb(Ea)(HEa)]$_2$(NO$_3$)$_2$	$P2_1/a$	9.605	8.494	12.440	90	90	99.9	15
[Pb$_5$Cl$_6$(Me$_2$Ea)$_4$]	$14_1/a$	14.010	14.010	17.729	90	90	90	16
[Pb$_3$Br(Me$_2$Ea)$_3$]Br$_2$	$C2/c$	28.573	9.249	19.104	90	107.7	90	17
[Pb$_3$(Me$_2$Ea)$_3$I]I$_2$	$C2/c$	29.442	9.569	19.221	90	105.9	90	16
[Pb$_2$(Dea)](NO$_3$)$_2$·DMF	$P2_1/c$	13.361	13.220	9.410	90	91.2	90	18
[Pb$_2$(HDea)$_2$]Cl$_2$	$P\bar{1}$	6.932	12.247	9.739	79.6	98.3	101.2	18
(H$_3$Dea)$_2$[Pb$_2$Br$_6$]	$P2_1/c$	10.665	29.638	7.631	90	76.0	90	18
[Cu(HDea)(H$_2$O)]·(NO$_3$)$_2$·H$_2$O	$P4_2/n$	18.024	18.024	20.772	90	90	90	19
[Pb(H$_3$Tea)$_2$](Ac)$_2$	$C2/c$	10.373	16.079	13.707	90	99.3	90	20
[Pb(H$_3$Tea)$_2$](NO$_3$)$_2$	$C2$	17.913	8.681	7.080	90	104.0	90	21
[Pb(HTea)(NO$_3$)$_2$·(DMSO)]	$P\bar{1}$	9.046	10.165	10.664	91.4	101.1	110.2	21
[Pb$_2$(HTea)(Ac)$_2$]	$P2_1/c$	8.754	20.856	9.138	90	106.9	90	21
[Pb$_2$(HTea)(SCN)]·(SCN)	$P2_1/n$	7.769	18.552	10.497	90	108.5	90	21
[Pb$_2$(HTea)]I$_2$	$Pna2_1$	18.555	17.914	8.119	90	90	90	22
[Cu(En)$_2$(HEa)]I$_2$	$P2_1/n$	11.401	10.806	12.925	90	90	87.4	23
Zn(En)$_2$(HEa)]I$_2$·0.5En	$A2/a$	31.87	9.451	12.116	90	90	96.7	24

dral environments. The ionic complex [Cu(Ea)(HEa)$_2$][Cu(Ea)(SCN)$_2$] [13] has copper atoms in square (cation) and tetrahedral (anion) configurations. Three types of coordination polyhedron (trigonal and square pyramids with lead atoms at apices, and bisphenoid with the lone pair of electrons in the equatorial plane) are present in structures [Pb$_3$(Me$_2$Ea)X]X$_3$, X = Br, I [16,17]. [Pb$_6$(Ea)$_6$I(HEa)$_2$]I$_5$ is built from two nonequivalent octahedrons, three five-apical and one seven-apical ones [14].

The acidic ligands may have two different positions, being coordinated and noncoordinated in the same molecule, as well as different modes of coordination. Thus, complexes [PbEnI]I·DMSO [5], [Pb$_6$(Ea)$_6$I(Hea)$_2$]I$_5$ [14], [Pb$_3$(Me$_2$Ea)X]X$_2$ (Fig. 5.1, A) and [Pb$_2$(HTea)(SCN)](SCN) [21] contain both coordinated and ionic bromide, iodide and thiocyanate groups. Complexes [Pb(Ac)$_2$(En)] (Fig. 5.1, B) and [Cu(Ea)(HEa)$_2$][Cu(Ea)(SCN)$_2$] (Fig. 5.1, C) contain cyclic and bridging acetic and thiocyanate groups. The (H$_3$Dea)$_2$[Pb$_2$Br$_6$] complex (Fig. 5.1, D) has bridging and monodentate bromide ions. The Pb–Br bond lengths are 2.754–3.387 Å, and those of Pb–I are 3.087–3.627 Å. Organic ligands are coordinated distinctively too. Thus, [Pb(En)$_2$]Br$_2$ (Fig. 5.2, A) contains bridging and cyclic ethylenediamine, and [Zn(En)$_2$(HEa)]I$_2$·0.5En has cyclic and noncoordinated En [24]. Complexes of monoethanolamine often contain both neutral and deprotonized aminoalcohol groups in one molecule. The [Cu(Ea)(HEa)$_2$][Cu(Ea)(SCN)$_2$] complex (Fig. 5.2, B) contains both mono- and bidentate aminoethoxy groups, and [Pb$_6$(Ea)$_6$I(HEa)$_2$]I$_5$ [14] contains these groups in tri- (μ_2-bridge) and tetra- (μ_3-bridge) dentate coordination. 2-Dimethylaminoethoxy groups in [Pb$_3$(Me$_2$Ea)X]X$_2$ (X = Br, I) compounds (Fig.5.1, A) are coordinated in the same way. Complexes [Pb$_2$(HTea)(Ac)$_2$] and [Pb$_2$(HTea)(SCN)](SCN) and [Pb$_2$(NO$_3$)$_2$(HTea)(DMSO)], (Fig. 5.2, C,D,E) contain doubly deprotonized triethanolamine with one noncoordinated oxygen atom and the other two are μ_2- and μ_3-bridges in tri- and tetra-coordination, respectively.

Bridging groups favour the formation of polynuclear and polymeric structures. Thus, [Cu(Ea)(HEa)$_2$][Cu(Ea)(SCN)$_2$] (Fig. 5.1, C) contains a dimeric anion. The [Pb(Ea)(HEa)]$_2$(NO$_3$)$_2$ complex (Fig. 5.3, B) contains a dimeric cation. The structurally independent part of the [Pb$_6$(Ea)$_6$I(HEa)$_2$]I$_5$ complex (Fig. 5.3, A) consists of three associated structures: two centrosymmetrical dimers and a tetramer with a pseudocubane structure [14]. Dimeric centrosymmetrical four-nucleus cations are found in the complexes [Pb$_2$(Dea)] (NO$_3$)$_2$·DMF and Pb$_2$(HDea)$_2$]Cl$_2$ (Fig. 5.4, A

(A)

(B)

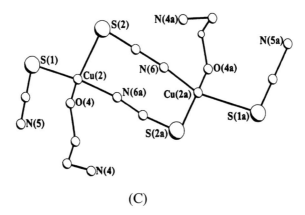

(C)

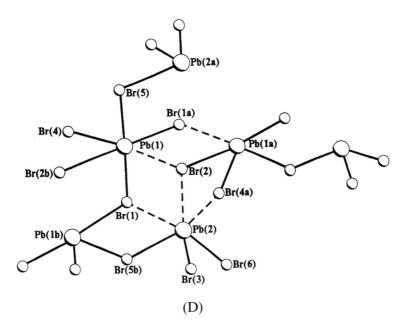

(D)

Fig. 5.1. Crystal structures of the complexes: A, [Pb$_3$(Me$_2$Ea)$_3$I]I$_2$; B, [Pb(Ac)$_2$(En)]; C, [Cu(Ea)(HEa)$_2$][Cu(Ea)(SCN)$_2$]; D, (H$_3$Dea)[Pb$_2$Br$_6$].

(A)

(B)

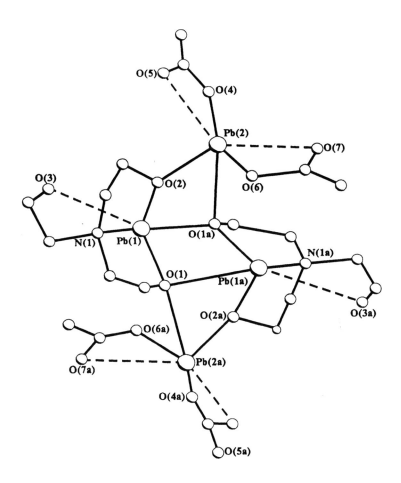

(C)

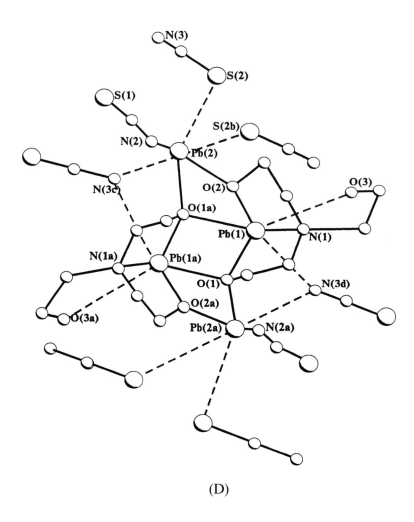

(D)

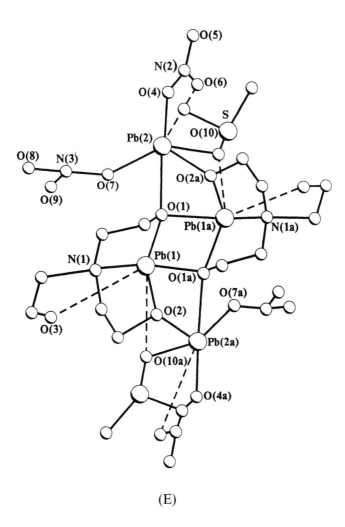

(E)

Fig. 5.2. Crystal structure of the complexes: A, [Pb(En)$_2$]Br$_2$; B, [Cu(Ea)(HEa)$_2$][Cu(Ea)(SCN)$_2$]; C, [Pb$_2$(HTea)(Ac)$_2$]; D, [Pb$_2$(HTea)(SCN)](SCN); E, [Pb$_2$(HTea)(NO$_3$)$_2$(DMSO)].

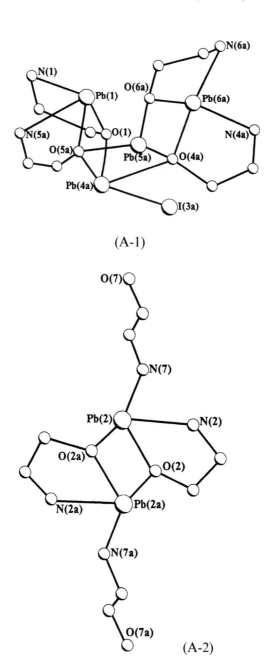

(A-1)

(A-2)

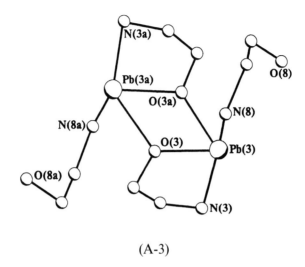

(A-3)

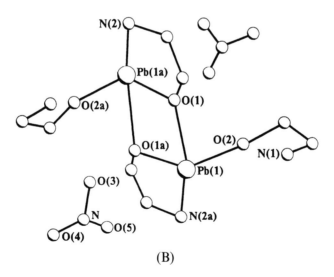

(B)

Fig. 5.3. Crystal structure of the complexes: A, $[Pb_6(Ea)_6(HEa)]I_5$ (three associated structures); B, $[Pb(Ea)(HEa)]_2(NO_3)_2$.

(A)

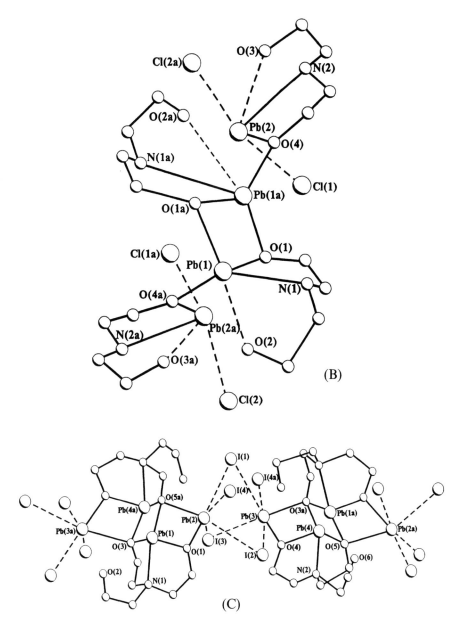

Fig. 5.4. Crystal structure of the complexes: A, $[Pb_2(Dea)](NO_3)_2 \cdot DMF$; B, $[Pb_2(HDea)_2]Cl_2$; C, $[Pb_2(HTea)]I_2$

and B). Four-nucleus cations are present in the compounds $[Pb_2(HTea)]I_2$ (Fig. 5.4, C), and $[Pb_2(HTea)(Ac)_2]$, $[Pb_2(HTea)(SCN)](SCN)$ and $[Pb_2(NO_3)_2(HTea)(DMSO)]$ (Fig. 5.2, C–E). The last three complexes are centrosymmetrical and contain lead–oxygen skeletons with a "ladder" structure [25]. The central flat fragment of the complexes is rhombic Pb_2O_2 [21]. Four atoms of lead form a parallelogram with a short side of 3.764–3.843 Å and a short diagonal of 3.823–3.894 Å. These values are less than double the van der Waals radius of lead and may indicate some specific interaction between the metal atoms.

Compounds $[Pb(Ac)_2(En)]$ (Fig. 5.1, B) and $[Pb(En)_2]Br_2$ (Fig. 5.2, A) are polymeric with a bridging function of acetic groups or molecules of ethylenediamine.

Only one complex, $[Pb_2(NO_3)_2(HTea)(DMSO)]$ (Fig. 5.2, E) contains coordinated solvent. In other cases, DMF or DMSO molecules are bound by hydrogen bonds or by van der Waals interaction. The $[Cu(HDea)(H_2O)](NO_3)\cdot 2H_2O$ complex contains both coordinated and noncoordinated water molecules [19].

The crystal structure of nickel complexes [8,10,11,26] consists, as a rule, of islands of coordinated cations or molecules with the central atom in a distorted octahedral environment. The Ni–N and Ni–O bond lengths in the equatorial plane are 2.02–2.09 Å. The axial lengths, meanwhile, are 2.3 Å. The angles of the equatorial bonds are 80–90° and the axial angles are 180°. The difference between the equatorial and axial lengths is considerable in the ethylenediamine complexes due to the peculiarity of the ligand steric structure.

The nearest-coordination environment of the copper atom in $[Cu(Ea)(HEa)_2]I$ [12] and $[Cu(En)_2]I_2\cdot 0.33DMF$ [1] is pseudo-square-planar, distorted tetrahedral (for example Cu(2) in $[Cu(Ea)(HEa)_2][Cu(Ea)(SCN)_2]$ [13]), and distorted square-pyramidal ($[Cu(En)_2(HEa)]I_2$ [23] or Cu(1) in the $[Cu(Ea)(HEa)_2][Cu(Ea)(SCN)_2]$ complex [13]) configurations. Equatorial interatomic distances are 1.91–2.04 Å for Cu–O and 2.01–2.10 Å for Cu–N. In complexes with aminoalcohols [12,13,19] the hydroxy-oxygen atoms are formally included into the coordination, converting the square-planar configuration into a distorted octahedron. The Cu–O distances are 0.5 Å ($\sim 25\%$) greater than the average length of the equatorial bonds. Therefore it is not possible to suppose formation of covalent bond.

Coordination compounds of zinc [6,9,24] are mononuclear and are built of complex islands. Zinc atoms have a tetrahedral or octahedral environment. Zn–N distances are 1.96–2.20 Å, and those of Zn–O are 2.23 Å.

The coordination environment of the lead atom is a multilayer type [2–4,14–18,20,21]. As a rule, the structural unit of the complex is a trigonal pyramid PbA_3 (A is a donor atom of oxygen or nitrogen). The lead atom forms three short covalent bonds (2.20–2.40 Å). Additional contacts of lead with donor atoms at greater distances (2.40–2.60 Å) transform the trigonal pyramid into a distorted pseudo-trigonal bipyramid, distorted square pyramid or distorted octahedron or dodecahedron. Complexes of molecular type have the oxygen atoms of hydroxy groups at distances of 2.60–2.80 Å forming the third level in aminoalcohol complexes. In ionic complexes, the last level is formed by the anions (Cl, Br, I, NO_3). The anions and hydroxyl oxygens have their positions in the region of most probable localization of the lone pair of lead electrons. It is these sterically active unpaired electrons in the hybrid orbital of the lead atom, as well as the space structure of the polydentate ligand, that determine the unsymmetrical and multilayer coordination environment of the central atom. If lone pairs of electrons are located in the s-orbital of the isotropic valence level, as for Pb(1) in the $[Pb_5Cl_6(Me_2Ea)_4]$ [16], $[Pb(H_3Tea)_2](NO_3)_2$ [21] and $[Pb(H_3Tea)_2](Ac)_2$ [20] complexes, the environment becomes relatively homogeneous with a coordination number of 8 in the first case, and 6 in the last two.

Therefore, the formation of the coordination sphere of the metal during complex formation in the conditions of direct synthesis does make it possible to obtain compounds with unusual spatial structures.

REFERENCES

1. Vassilyeva, O.Yu.; Kokozay, V.N.; Simonov, Yu.A. *Zh. Neorg. Khim.* **36**(12), 119 (1991).
2. Polyakov, V.R.; Kokozay, V.N. *J. Coord. Chem.* **32**, 343 (1994).
3. Kokozay, V.N.; Polyakov, V.R.; Dvorkin, A.A. *Zh. Neorg. Khim.* **37**(1), 64 (1992).
4. Kokozay, V.N.; Sienkiewicz, A.V. *Polyhedron* **13**(9), 1427 (1994).
5. Kokozay, V.N.; Polyakov, V.R.; Simonov, Yu.A. *J. Coord. Chem.* **28**, 191 (1993).
6. Dvorkin, A.A.; Kokozay, V.N.; Petrusenko, S.R.; Sienkiewicz, A.V. *Dokl. Akad. Nauk Ukr. SSR. Ser. B.* No. 10, 31 (1989).
7. Sobolev, A.N.; Shvelashvili, A.E.; Kokozay, V.N.; Petrusenko, S.R.; Kapshuk, A.A. *Soobshch. Akad. Nauk GSSR.* **136**(2), 325 (1989).

8. Dvorkin, A.A.; Kokozay, V.N.; Petrusenko, S.R.; Simonov, Yu.A. *Ukr. Khim. Zh.* **57**(1), 5 (1991).

9. Kokozay, V.N.; Petrusenko, S.R.; Sienkiewicz, A.V.; Simonov, Yu.A. *Ukr. Khim. Zh.* **59**(3), 236 (1993).

10. Kokozay, V.N.; Simonov, Yu.A.; Petrusenko, S.R.; Dvorkin, A.A. *Zh. Neorg. Khim.* **35**(11), 2839 (1990).

11. Simonov, Yu.A.; Skopenko, V.V.; Kokozay, V.N.; Petrusenko, S.R.; Dvorkin, A.A. *Zh. Neorg. Khim.* **36**(3), 625 (1991).

12. Vassilyeva, O.Yu.; Kokozay, V.N.; Simonov, Yu.A. *Dokl. Akad. Nauk Ukr. SSR. Ser. B* No. 10, 138 (1991).

13. Vassilyeva, O.Yu.; Kokozay, V.N.; Skopenko, V.V. *Ukr. Khim. Zh.* **60**(3–4), 227 (1994).

14. Skopenko, V.V.; Kokozay, V.N.; Polyakov, V.R.; Sienkiewicz, A.V. *Polyhedron* **13**(1), 15 (1994).

15. Kokozay, V.N.; Polyakov, V.R.; Simonov, Yu.A.; Sienkiewicz, A.V. *Zh. Neorg. Khim.* **37**(8), 1810 (1992).

16. Kokozay, V.N.; Sienkiewicz, A.V. *Polyhedron* **13**(9), 1439 (1994).

17. Kokozay, V.N.; Sienkiewicz, A.V. *J. Coord. Chem.* **31**, 1 (1994).

18. Sienkiewicz, A.V.; Kokozay, V.N. *Z. Naturforsch. Teil B* **4b**, 615 (1994).

19. Ivanova, E.I.; Kokozay, V.N. *Dokl. Akad. Nauk Ukr. SSR. Ser. B* No. 9, 135 (1994).

20. Kokozay, V.N.; Sienkiewicz, A.V. *J. Coord. Chem.* **30**(3–4), 245 (1993).

21. Sienkiewicz, A.V.; Kokozay, V.N. *Polyhedron* **13**(9), 1431 (1994).

22. Kokozay, V.N.; Sienkiewicz, A.V. *Polyhedron* **12**(19), 2421 (1993).

23. Kokozay, V.N.; Dvorkin, A.A.; Vassilyeva, O.Yu.; Sienkiewicz, A.V.; Rebrova, O.N. *Zh. Neorg. Khim.* **36**(6), 1446 (1991).

24. Kokozay, V.N.; Petrusenko, S.R.; Simonov, Yu.A. *Zh. Neorg. Khim.* **39**(1), 81 (1994).

25. *Comprehensive Coordination Chemistry. The Synthesis, Reactions, Properties*, Vols 3–5 (Ed.: Wilkinson, G.). Oxford: Pergamon Press (1987).

26. Kokozay, V.N.; Petrusenko, S.R. *Ukr. Khim. Zh.* **61**(3), 3 (1995).

Author Index

Subject Index

A

Acetophenone 98
Acetylacetonates 59, 93, 114, 115
Acetylacetone 59, 98
Acetylene 5, 47, 49–51
2-Acetylpyrrole 91
Acidoligands 220
Acids
  acetic 88
  benzoic 88
  isobutyric 88
  oxalic 88
  phthalic 88
  propionic 88
  salicylic 88
  succinic 88
Active centers 136, 138, 142
  thread-like branching 139
Active magnesium 53, 56
Active proton 163, 165
Acyl halides 5, 52
Adamantane 16, 22, 48
Adducts 82, 85
  bimetallic 91, 124
L-Alanine 88
Alcoholates 51
Alcohols 51
Alkanedithiols 87, 102
Alkanes 11
Alkynes 1, 47–50
Allene 16, 21, 48
Allyl bromide 53
Allyl sulfide 118
Aluminum carbonyl 9
Amine basicity 181
Aminoalcohols 182
4-Aminoantipyrine 109
2-(2-Aminoethyl)pyridine 108

2-[(2-Aminoethyl)thiomethyl]-
  benzimidazole 98
Ammines 170
Amminosolvates 170
Ammonia 11, 155, 170
Ammonium salts 77, 83, 169–171
Aniline 32
Anion coordination 182
Antipyrine fragment 95
Arenes 27, 33, 34, 35
Aromatic compounds 1
Autocatalytic mechanism 141
Azines and bis-azines 39, 102
Azoles 100, 107
Azomethines 83, 91, 94, 95, 98, 104, 110,
  204, 209, 214

B

Bacterial dissolution of metals 168
Beckman regrouping 165
N-Benzacylsalicylamide 211
Benzene 5, 24, 36, 40, 41, 45, 46, 85, 86, 88,
  120
N-Benzylacylsalicylamide 211
N-Benzylidenacylsalicylamide 211
Benzimidazole-2-aldehyde 98
Benzimidazoles 98
1,7-bis(2-Benzimidazolyl)-2,6-
  dithiahexane 107
Benzoylacetone 109
Benzyl chloride 88
Bicyclo[2.2.1]heptene 20
Bipyridyl 39, 47, 63, 82, 83, 85–88, 103, 158
Bisphenoid 220
Bonding
  in alkene complexes 16, 17
  in CO complexes 10
Boron trichloride 13